Programmable
Controllers

Programmable Controllers

Thomas A. Hughes

Resources for Measurement and Control Series

Instrument Society of America

INSTRUMENT SOCIETY OF AMERICA
67 Alexander Drive
P.O. Box 12277
Research Triangle Park
North Carolina 27709

Library of Congress Cataloging-in-Publication Data

Hughes, Thomas A.
  Programmable controllers / Thomas A. Hughes.
     p.   cm.—(Resources for measurement and control series)
  Includes bibliographical references.
  ISBN 1-55617-166-8
  1. Programmable controllers. I. Title. II. Series.
  TJ223.P76H84   1989
  629.8'95—dc20   89-39267
                                                          CIP

ISBN: 1-55617-166-8

To my wife Ellen, my daughter Audrey,
my Mother and Father,
and all my brothers and sisters
for their love and encouragement
over the years.

Contents

Preface

This book provides a comprehensive coverage of the fundamental principles of programmable controllers. It is written for engineers, technicians, and sales and management personnel who are new to process and machine automation and need to understand the basic principles of programmable controllers. It is also intended as an easy-to-read reference manual for experienced personnel.

The basic principles of programmable controllers are discussed in Chapter 1. The second chapter covers numbering systems and binary codes used in computer systems. Chapter 3 discusses the basic concepts of logic gates, logic laws, digital devices, and logic design to provide a strong foundation in logic applications. The fourth chapter discusses the fundamentals of electricity, electrical conductors, the operations of relays and electrical solenoids, electrical ladder diagrams and symbols, and typical electrical control applications.

Chapter 5 covers the input/output systems used in programmable controller systems, and Chapter 6 discusses the memory and storage devices used in programmable controllers. The basic programming languages, ladder and Boolean, are covered in detail in Chapter 7. High-level languages, such as function block and sequential function chart, are discussed in Chapter 8. Chapter 9 provides a basic understanding of data communications terminology and concepts and their application to programmable controller-based systems. The tenth chapter covers the design and application of programmable controllers to industrial processes and machines; and the final chapter discusses equipment installation, system testing and start-up principles, and maintenance procedures required to obtain a reliable and maintainable control system.

Tom Hughes
Arvada, CO
May 1989

About the Author

Thomas A. Hughes, a senior member of the Instrument Society of America, has twenty years experience in the design and application of instrumentation and control systems to industrial processes, including ten years in the management of instrumentation projects.

He received his B.S. in engineering physics from the University of Colorado and his M.S. in electrical engineering from Colorado State University.

Mr. Hughes holds professional licences in the states of Colorado and Alaska and has held engineering positions with Dow Chemical, B. K. Sweeney Manufacturing, and Stearns-Roger Engineering. He has taught numerous courses in electronics, mathematics, and control systems at the college level and in industry and is currently Manager of Control Systems Development with Rockwell International in Golden, Colorado.

Programmable Controllers

1

Introduction to Programmable Controllers

Introduction Programmable controllers were originally designed to replace relay-based logic systems and solid-state hard-wired logic control panels. Their advantages over conventional logic systems are that they are easily programmed, highly reliable, flexible, relatively inexpensive, and able to communicate with other plant computers.

A programmable controller examines the status of inputs and, in response, controls some process or machine through outputs. Combinations of input and output data are referred to as logic. Several logic combinations are usually required to carry out a control plan or program. This control plan is stored in memory using a programming device to input the program into the system. The control plan in memory is periodically scanned by the processor, usually a microprocessor, in a predetermined sequential order. The period required to evaluate the programmable controller program is called the "scan time".

A simplified diagram of a programmable controller is shown in Figure 1-1, where instruments, such as limit switches and panel-mounted push buttons, are wired to input modules, and the output modules in turn drive devices such as electric solenoid valves and indicator lights. The input and output modules are controlled by the logic unit.

Figure 1-1 shows a typical configuration of the early programmable controller systems, which were intended to replace relay or hard-wired logic systems. As we will see in later discussions the modern programmable controller is far more complex and powerful.

Brief History of Programmable Controllers In 1968, a major automobile manufacturer wrote a design specification for the first programmable controller. The primary goal was to eliminate the high cost associated with the frequent replacement of inflexible relay-based control systems. The specification also called for a solid-state industrial computer that

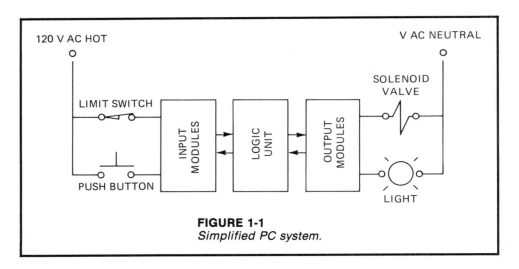

FIGURE 1-1
Simplified PC system.

could be easily programmed by maintenance technicians and plant engineers. It was hoped that the programmable controller would reduce production downtime and provide expandability for future production changes. In response to this design specification, several manufacturers developed logic control devices called programmable controllers.

The first programmable controller was installed in 1969, and it proved to be a vast improvement over relay-based panels. Because they were easy to install and program, they used less plant floor space and were more reliable than relay-based systems. The initial programmable controller not only met the automobile manufacturer production needs, but further design improvements in later models led to widespread use of programmable controllers in other industries.

There were probably two main factors in the initial design of programmable controllers that led to their success. First, highly reliable solid-state components were used, and the electronic circuits and modules were designed for the harsh industrial environment. The system modules were built to withstand electrical noise, moisture, oil, and the high temperatures encountered in industry. The second important factor was that the initial programming language selected was based on standard electrical ladder logic. Earlier computer systems failed because plant technicians and engineers were not easily trained in computer programming. However, most were trained in relay ladder design, so that programming in a language based on relay circuit design was learned quickly.

When microprocessors were added in 1974 and 1975, the basic capabilities of programmable controllers were greatly expanded and improved. They were able to perform sophisticated math and data manipulation functions.

In the late 1970s, improved communications components and circuits made it possible to place programmable controllers thousands of feet from the equipment they controlled, and several programmable controllers can now exchange data to more effectively control processes and machines. Also, microprocessor-based input and output modules allowed programmable controller systems to evolve into the analog control world.

Programmable controllers are found in thousands of industrial applications. They are used to control chemical processes and facilities. They are found in material transfer systems that transport both the raw materials and the finished products. They are used with robots to perform hazardous industrial operations to allow man to perform more intelligent functions. Programmable controllers are used in conjunction with other computers to perform process and machine data collection and reporting functions, including statistical process control, quality assurance, and diagnostics. They are utilized in energy management systems to reduce costs and improve environmental control of industrial facilities and office buildings.

Basic Components of a PC System
Regardless of size, cost or complexity, all programmable controllers share the same basic components and functional characteristics. A programmable controller will always consist of a processor, an input/output system, a memory unit, a programming language and device, and a power supply. A block diagram of a typical programmable controller system is shown in Figure 1-2.

The Processor   The processor consists of one or more standard or custom microprocessors and other integrated circuits that perform the computing and control functions of the PC system. Since the processor in most programmable controller systems is microprocessor-based, we will discuss the basic operation of microprocessors.

A microprocessor can be defined as a collection of digital circuit elements connected together to form an information processing unit. The five essential elements of a microprocessor are shown in Figure 1-3. The five elements are a system clock that synchronizes the responses of the system's components;

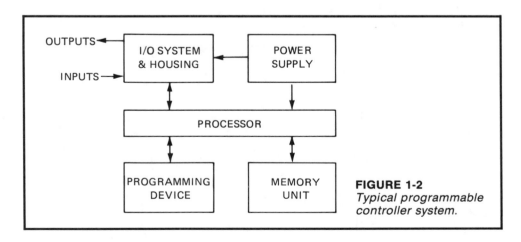

FIGURE 1-2
Typical programmable controller system.

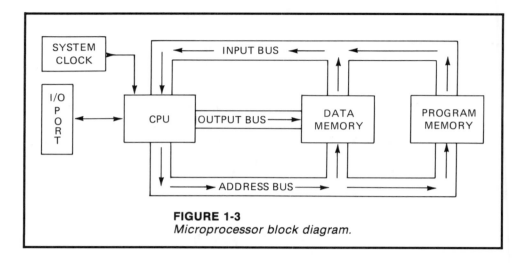

FIGURE 1-3
Microprocessor block diagram.

program memory that stores the program the system will execute; a data memory to store the numbers that are being manipulated; I/O ports for connecting to exterior devices; and a central processing unit (CPU), which operates on the data in the sequence determined by the program. For example, if the system is required to find the sum of a set of numbers, then the numbers will be stored in the data memory, the summing program is stored in the program memory, and the actual calculations will be done by the microprocessor.

The CPU is the focal point of the system, but it must have facilities for storing the control program, and there must be some sort of data storage. The program is stored in memory as a set of binary numbers (0 or 1), which are "coded" to represent the various steps the microprocessor must execute. The microprocessor contains digital circuits that can decode those program instructions and implement the prescribed program steps.

Inside the microprocessor are several digital circuits called "registers," which are used to store binary numbers of particular significance. The most important registers are as follows:

1. The accumulator: It is the central point for all data manipulation (i.e., add, subtract, shift, etc.).

2. The index register: It is used to store and create data addresses that are important in a software program.

3. The instruction register: In order to have the microprocessor system execute a particular program, the program memory sends out a series of commands or "instructions" to the CPU. As each instruction is received by the CPU, it is stored in the instruction register. The CPU then carries out the operation required before receiving the next instruction.

4. The program counter (PC): It is used to keep track of the CPU's progress through the program.

An important feature of a microprocessor is the manner in which the three system elements (CPU, program memory, and data memory) communicate with each other by means of buses. A bus is defined as a set of physical connections along which parallel binary information can be transmitted. Only one word of parallel information can exist on a bus at any given time. For example, in Figure 1-3, it would not be permissible for the CPU to simultaneously send an address to both the program and data memory via the address bus.

Usually the CPU assumes control of the various buses by means of special signals sent out on control lines to the data and program memories. Sometimes these control signals are grouped together under the heading "control bus," but for clarity the control signals are not shown in the block diagrams in this section.

In executing a program, the microprocessor uses the data buses to exchange information in a well-defined manner. The complete sequence of information exchange that carries out one program step is known as an "instruction cycle". An example instruction cycle might consist of the following steps:

1. Fetch the next instruction (i.e., program step) from program memory and place it in the instruction register.

2. Increment the program counter by one.

3. Execute the instruction.

To illustrate the operation of a simple computer instruction, we will look at the operation of a typical microprocessor when power is first applied. The system clock starts sending out pulses to the system circuits at the rate of several million pulses per second. Every action of the microprocessor is regulated precisely by this clock pulse, which is independent of the system's other control signals.

On the first clock pulse, a reset signal automatically sets all the CPU registers to zero, so that the program counter now contains all zeros (00000000). The system is now ready to carry out a process called bootstrapping or booting, so called because the microprocessor, in effect, pulls itself up "by its own bootstraps". At the next clock pulse the program counter is loaded with an address set by the system designer. The address is a sequence of high and low voltages (or 1s and 0s) that identifies the location in program memory of the bootstrap program. For simplicity, the address is shown in Figure 1-4 as only eight bits long; in reality, most microprocessors have addresses 16 bits long.

The CPU processes the start-up program in thousands of steps, each step consisting of eight binary bits (a byte). It is important to realize that all microprocessors are sequential machines that operate on 8, 16, or 32 binary bits, so when we talk about a program scan we are discussing millions of tiny steps. A byte may represent an address or the instruction or piece of data located at a given address. Each byte moves as a sequence of 1s and 0s on the address bus or on the data bus.

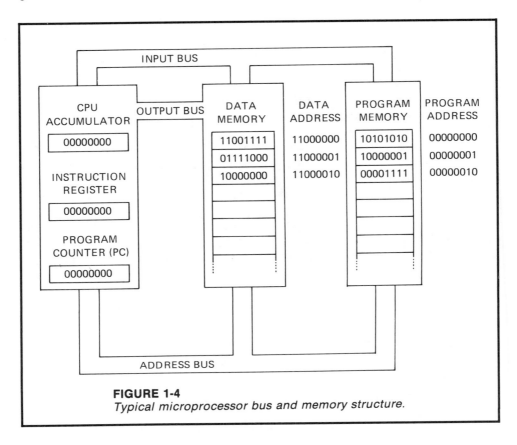

FIGURE 1-4
Typical microprocessor bus and memory structure.

After the first clock pulse sets all the bits in the program counter to zero, the next clock pulse loads a given address into the program counter as shown in Figure 1-5. On the next clock pulse, the CPU moves the address (00000111) from the program counter to the address bus. By the end of the pulse, the next address in the program appears in the program counter.

As the clock runs, circuitry inside program memory alerts the correct memory cell. As shown, the binary address string (00000111) is different from the string (01111110) representing the contents stored at the address. In this example, the eight-bit string in program memory is the binary encoding of the first instruction in the start-up program. The CPU will next read this "instruction" but must first wait for a control signal and the next clock pulse. At the next beat of the system clock, the CPU sends a read signal to program memory, which instantly transfers the data onto the input bus.

Once on the input bus, information selected from this initial address moves back to the CPU. On the next clock pulse, the CPU takes the byte of data from the input bus and sends it to its registers. Since this is the first data the CPU has received since power was turned on, it interprets the data as an instruction to be acted upon during the next clock pulse. This sequence of fetching the instruction and then performing the step in the program will be repeated thousands of times until all the instructions in the boot program are executed.

After the boot-up program is complete, the system will run a selected program or wait for input instructions from a keyboard or other input device.

The processor uses the input/output (I/O) port to communicate with the outside. However, since the processor uses 0 to 5 V dc only, there must be an I/O system to convert the processor logic signals to the numerous signals encountered in process control.

Input/Output System   The input/output (I/O) system provides the physical connection between the process equipment and the processor. The programmable controller system uses various input circuits or modules to sense and measure physical quantities of the process, such as motion, level, temperature, pressure, position, current, and voltage. Based on the status sensed or values measured, the processor controls various output modules to drive field devices such as valves, motors, pumps, and alarms to exercise control over a machine or a process.

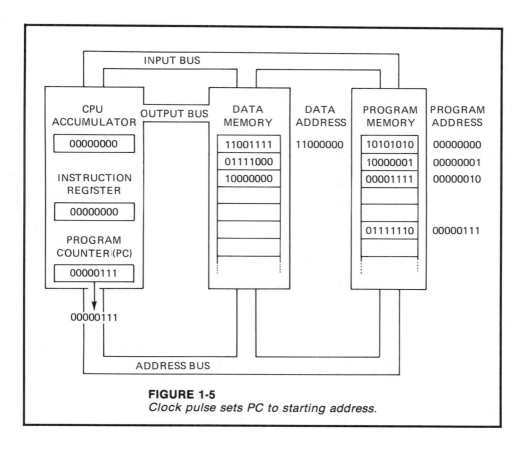

FIGURE 1-5
Clock pulse sets PC to starting address.

Inputs

The inputs from instruments or sensors supply the data and information the processor needs to make logical decisions. These input signals from varied devices such as push buttons, hand switches, thermocouples, strain gages, etc., are connected to input modules to filter and condition the signal for use by the processor.

Outputs

The outputs from the programmable controller energize or deenergize control devices to regulate processes or machines. These output signals are control voltages from the output modules, and they are generally not power voltages. For example, an output module sends a control signal that energizes the coil in a motor starter. The energized coil closes the power contacts of the starter. These contacts then close to start the motor. The output modules are usually not directly connected to the power circuit but rather to devices such as motor starter and heater contactors that apply power circuit voltages to the control devices.

Modular Housings

One programmable controller manufacturer features over 100 discrete, analog, and intelligent I/O modules. These modules are mounted in "universal" modular housings as shown in Figure 1-6. The term "universal" in this context means that any module can be inserted into any I/O slot.

Modular I/O housings are designed so that the I/O modules can be removed without turning off the ac power or removing the field wiring. Most I/O modules are mounted on printed circuit boards that can be inserted into an I/O housing or a card rack. The backplane of the housings into which the modules are plugged have a printed circuit card that contains the parallel communications bus to the processor and the dc voltages to operate the logic circuits in the I/O modules.

These I/O housings can be mounted in a control panel or on a sub-panel in an enclosure. The housings are designed to protect the I/O module circuits from dirt, dust, electrical noise, and mechanical vibration.

Housings generally come in three sizes with 4, 8, or 16 slots, as shown in Figure 1-6. The backplane of the I/O chassis has sockets for each module. These sockets provide the power and data communications connection to the processor for each module. The modules are designed for a wide range of manufacturing and process applications and include the following I/O types: discrete, analog, digital, and intelligent.

Discrete Inputs/Outputs

Discrete is the most common class of input/output in a programmable controller system. This type of interface module connects field devices that have one of two states, such as on/off or open/closed, to the processor. Each discrete I/O module is designed to be activated by some field-supplied voltage signal, such as 220 V ac, 120 V ac, and so on.

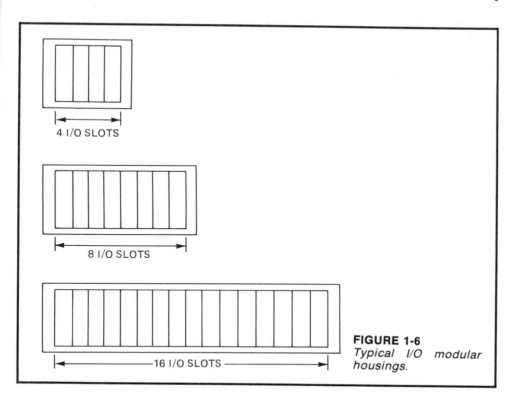

FIGURE 1-6
Typical I/O modular housings.

4 I/O SLOTS

8 I/O SLOTS

16 I/O SLOTS

In a discrete input module, if an input switch is closed, the input interface module senses the supplied voltage and converts it to a logic-level signal acceptable to the processor to indicate the status of that device. A logic 1 indicates ON or CLOSED, and a logic 0 indicates OFF or OPENED.

A typical discrete input module is shown in Figure 1-7. Most input modules will have a light-emitting diode (LED) to indicate the status of each input.

In a discrete output module, the output interface circuit switches the supplied control voltage that will energize or deenergize the field device. If an output is turned ON through the control program, the supplied control voltage is switched by the interface circuit to activate the referenced (addressed) output device.

Figure 1-8 shows a typical discrete output module. It can be thought of as a simple switch through which power can be provided to control the output device. During normal operation, the processor sends the output state determined by the logic program to the output module. The module then switches the power to the field device.

An electric fuse is normally provided in the output circuit of the module to prevent excessive current from damaging the output module. If the fuse is not provided in the output module, it should be provided in the system design.

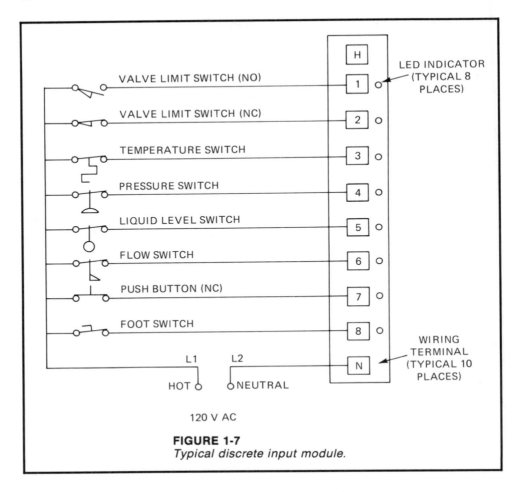

FIGURE 1-7
Typical discrete input module.

Most output modules use opto-isolators to separate field voltages from the logic circuits in the module.

Analog I/O Modules

The analog I/O modules allow for monitoring and controlling of analog voltages and currents, which are compatible with many sensors, motor drives, and process instruments. With the use of analog I/O, most process variables can be measured or controlled with appropriate interfacing.

Analog I/O interfaces are generally available for several standard unipolar (single polarity) and bipolar (negative and positive polarity) ratings. In most cases, a single input or output interface can accommodate two or more different ratings and can satisfy either a current or voltage requirement. The different ratings will be either hardware- (switches or jumpers) or software-selectable.

Digital I/O Modules

Digital I/O modules are similar to discrete I/O modules in that discrete ON/OFF signals are processed. However, the main difference is that discrete I/O interfaces require only a single bit to read an input or control an output. On the other hand, digital I/O modules process a group of discrete bits in parallel form or in serial form.

Typical devices that interface with digital input modules are binary encoders, bar code readers, and thumbwheel switches. Some instruments driven by digital output modules include LED displays, intelligent panels, and BCD displays.

Intelligent I/O Modules

The discrete, analog, and digital I/O modules will normally cover 90% of the I/O applications encountered in programmable controller systems. However,

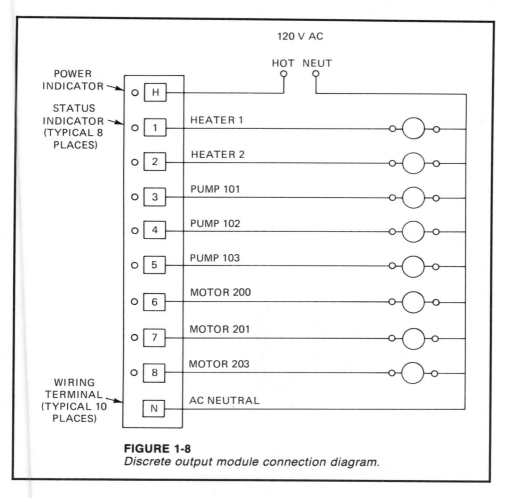

FIGURE 1-8
Discrete output module connection diagram.

to process certain types of signals or data efficiently, the programmable controller system will require special modules. These special interfaces include those that condition input signals, such as thermocouple modules, or other signals that cannot be interfaced using standard I/O modules. Special I/O modules may also use an on-board microprocessor to add intelligence to the interface. These intelligent modules can perform complete processing functions independent of the CPU and the control program scan.

A typical set of intelligent I/O modules is shown in Figure 1-9.

Memory   Memory is used to store the control program for the PC system; it is usually located in the same housing as the CPU. The information stored in memory determines how the input and output data will be processed.

Memory elements store individual pieces of data called bits. A bit has two states: 1 or 0, on or off, true or false, etc. Memory units are mounted on circuit boards and are usually specified in thousands or "K" increments where 1K is 1024 words (i.e., $2^{10} = 1024$) of storage space. Programmable controller memory capacity may vary from less than one thousand words to over 64,000 words (64K words) depending on the programmable controller manufacturer. The complexity of the control plan will determine the amount of memory required.

The programs and data stored in memory are generally described using four terms: (1) Executive, (2) Scratch Pad, (3) Control Program, and (4) Data Table. The *Executive* is a permanently stored collection of programs that is the operating system for the programmable controller. This operating system directs activities such as execution of the control program, communication with the peripheral devices, and other system housekeeping functions. The *Scratch Pad* is a temporary storage area used by the CPU to store a relatively small amount of data for interim control or calculations. Data that is needed quickly is stored in this area to increase the speed of data manipulation. The *Control Program* area provides storage for any programmed instructions entered by the user. The *Data Table* stores any data associated with the control program,

FIGURE 1-9
Typical intelligent I/O modules. (Courtesy of Allen-Bradley Co.)

such as timer/counter preset values and any other stored constants or variables that are used by the main program or the CPU. It also retains the input/output status information.

The storage and retrieval requirements are not the same for the executive, the scratch pad, the control program, and the data table; therefore, they are not always stored in the same type of memory. For example, the executive requires a memory that permanently stores its contents and cannot be deliberately or accidentally altered by loss of electrical power or by the user. On the other hand, the user would need to alter the control program and/or the data table for any given application.

Although there are several different types of computer memory, they can always be classified as *volatile* or *nonvolatile*. Volatile memory will lose its programmed contents if all operating power is lost or removed. Volatile memory is easily altered and quite suitable for most programming applications when supported by battery backup and/or a recorded copy of the program.

Nonvolatile memory will retain its data and program even if there is a complete loss of operating power. It does not require a backup system.

Programming Languages   The programming language allows the user to communicate with the programmable controller via a programming device or panel. Programmable controller manufacturers use several different programming languages, but they all convey to the system, by means of instructions, a basic control plan.

A control plan or program is defined as a set of instructions that are arranged in a logical sequence to control the actions of a process or machine. For example, the program might direct the programmable controller to turn on a motor starter when a push button is depressed and at same time direct the programmable controller to turn on a control panel-mounted RUN light when the motor starter auxiliary contacts are closed.

A program is written by combining instructions in a certain order. Rules govern the manner in which instructions are combined and the actual form of the instructions. These rules and instructions combine to form a language.

The four most common types of languages encountered in programmable controllers are as follows:

1. Ladder diagram
2. Boolean logic
3. Function blocks
4. Sequential function chart

Ladder Diagram

The most common language used is ladder logic. The reason for this is relatively simple. The original programmable controllers were designed to replace electrical relay-based control systems. These systems were designed by technicians and engineers using a symbolic language called ladder diagrams. The ladder diagram consists of a series of symbols interconnected by lines to indicate the flow of current through the various devices. The ladder drawing consists of basically two things: first is the power source, which forms the sides of the ladder (rails), and second is the current that flows through the various logic input devices that form the rungs of the ladder.

In electrical design, the ladder diagram is intended to show only the circuitry necessary for the basic operation of the control system. Another diagram, called the wiring diagram, is used to show the physical connection of control devices. The discrete I/O module diagrams shown earlier are examples of wiring diagrams. A typical electrical ladder diagram is shown in Figure 1-10. In this diagram a push button (PB1) is used to energize a pump start relay (CR1) if the level in a liquid storage tank is not high. Each device has a special symbol assigned to it to make reading of the diagram easier and faster. These symbols are discussed in detail in a later chapter on electrical design.

The same control application can be implemented using the programmable controller ladder diagram program as shown in Figure 1-11. The diagrams are read in the same manner from left to right, with the logic input conditions on the left and the logical outputs on the right. In the case of electrical diagrams there must be electrical continuity to energize the output devices; for programmable controller ladder programs there must be logic continuity to energize the outputs.

In ladder programs, three basic symbols or instructions are used to form the program. The first symbol is the same normally open relay contact symbol used in electrical ladder diagrams; this instruction uses the symbol (⊣ ⊢) in ladder programs. It instructs the processor to examine its assigned bit location in memory. If the bit is ON (logic 1), the instruction is true and there is logic continuity through the instruction on the ladder rung. If the bit is OFF (logic 0), there is no logic continuity through the instruction on the rung.

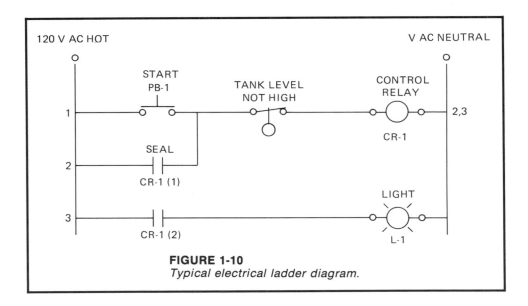

FIGURE 1-10
Typical electrical ladder diagram.

The second important symbol (⊣∕⊢) represents a normally closed instruction. Unlike the normally open instruction, it directs the processor to examine the bit for logical 0 or the OFF condition. If the bit is OFF, the instruction is true and there is logic continuity through the instruction. If the bit is ON, the normally closed instruction is false and there is no logic continuity.

The third important symbol is the output coil (—( )—) instruction. This instruction directs the processor to set a certain location in memory to ON or 1, if all the instructions in the logic path preceding it in the rung are true. If there is no complete logic path in the ladder rung, the processor sets the output coil bit to 0 or OFF.

Two other symbols are needed to write a complete ladder program. They are the branch start (⊢) and the branch end (⊤) instructions. These instructions are used to start and end parallel paths of logic to allow for OR functions in programs.

In Figure 1-11 the five-digit numbers above the instructions are the reference addresses. The reference address indicates where in the memory the logic operation will take place. In Figure 1-11, the normally open instruction for the start PB directs the processor to see if the reference address 110/00 is ON. In the same manner, the normally closed instruction for the tank level not high instructs the processor to see if the reference address 110/01 is OFF. If there is logic continuity through both instructions, the output coil at address 010/00 is turned ON. This bit is then used to "seal in" the start push button instruction and also turn on the output coil bit 010/01 to energize the pump run light.

Programming Devices   The programming device is used to enter, store, and monitor the programmable controller software. Programming devices can be dedicated or personal computers that normally have four basic components: keyboard, visual display, microprocessor, and communications cable.

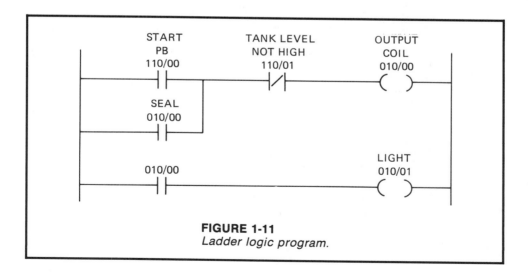

FIGURE 1-11
Ladder logic program.

The programming terminal is normally connected only to the programmable controller system during programming or during troubleshooting of the control system. Otherwise the programming device is disconnected from the system.

The most common programming terminals are as follows:

1. Hand-held manual programmer
2. Industrial programming terminal
3. Personal computer-based programmer

The hand-held programmers are inexpensive and portable units, normally used to program small programmable controllers. Most of these units resemble portable calculators, but with larger displays and a somewhat different keyboard. The displays are generally LED (light-emitting diode) or dot matrix LCD (liquid crystal display), and the keyboard consists of alphanumeric keys, programming instruction keys, and special function keys. Even though they are mainly used for inputting and editing the control program, the portable programmers are also used for testing, changing, and monitoring the program. Figure 1-12 shows a typical hand-held programmer.

The industrial programming terminal shown in Figure 1-13 is an intelligent device that not only displays the control program but also provides program editing functions independent of the programmable controller. The industrial programmer normally has a cathode ray tube (CRT) display and its own internal memory and program to create, alter, and monitor programs. The

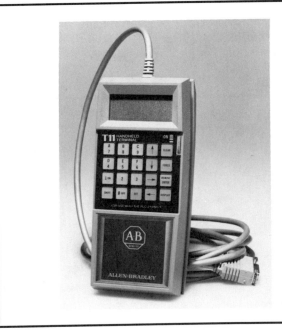

FIGURE 1-12
Typical hand-held programmer. (Courtesy of Allen-Bradley Co.)

FIGURE 1-13
Industrial programming terminal. (Courtesy of Allen-Bradley Co.)

CRT is a powerful tool for programming, since the control program can be edited and viewed without being connected to a programmable controller.

The latest innovation in programming terminals is the IBM® compatible personal computer types. A typical example of this type of terminal is shown in Figure 1-14. These units have all the features of industrial terminals such as program editing and storage. But, they also have added features such as automatic program printouts and connection to local area networks (LANs). LANs give the programmer or engineer access to any programmable controller in the network, so that any device in the network can be monitored and controlled.

Power Supply   The power supply converts ac line voltages to dc voltages to power the electronic circuits in a programmable controller system. These power supplies rectify, filter, and regulate voltages and currents to supply the correct amounts of voltage and current to the system. The power supply normally converts 120 V ac or 240 V ac line voltage into direct current voltages such as $+5$ or ± 15 V.

The power supply for a programmable controller system may be integrated with the processor, memory, and I/O modules into a single housing, or it might be a separate unit connected to the system through a cable. As a system expands to include more I/O modules or special function modules, most programmable controllers require an additional or auxiliary power supply to meet the increased power demand. Programmable controller power supplies are usually designed to eliminate electrical noise present on the ac power and signal lines of industrial plants so that this electrical noise does not introduce

FIGURE 1-14
Personal computer-based programming terminal. (Courtesy of Allen-Bradley Co.)

FIGURE 1-15
Typical programmable controller power supply. (Courtesy of Allen-Bradley Co.)

errors in the control system. They are also normally designed for the higher temperature and humidity present in most industrial environments. A typical power supply used in programmable controller systems is shown in Figure 1-15.

Advances in Programmable Controllers
The introduction of the low-cost personal computer in the early 1980s led to their use in programmable controller-based systems to generate process color graphics display. These process display screens can be easily modified for process changes, and the computer can perform other functions such as alarm listing, report generation, and programmable controller programming.

A typical programmable controller system with process graphics is shown in Figure 1-16. In this system, the personal computer is normally connected to the programmable controller rack using serial communications such as RS-232D. The communications module can be mounted in the I/O rack or outside, depending on the application.

The software for vendor-supplied process color displays is normally menu-driven and relatively easy to use. The process screens are normally based on the process and instrument drawings for the process being controlled.

Some programmable controller logic software packages run on a personal

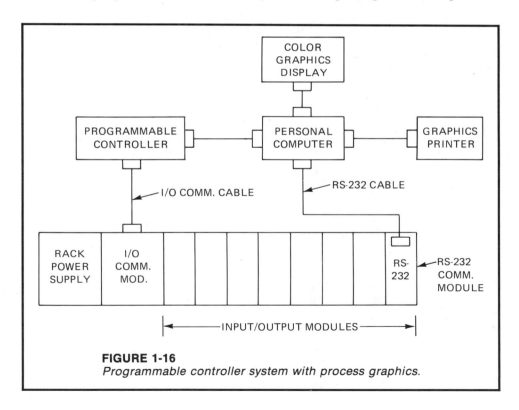

FIGURE 1-16
Programmable controller system with process graphics.

computer; therefore, the graphics computer with a printer can also be used to program as well as document the control system software.

EXERCISES

1.1 Explain the operation and purpose of each component in a typical programmable controller system.

1.2 List some typical examples of analog input and output signals found in process industries.

1.3 Using block diagrams of a typical microprocessor, explain the operation of a microprocessor performing the following program steps: (1) Fetch data (11110000) in memory location 00000001, (2) add to data (00100000) in memory location 00000010, and (3) store result in memory location 00000011.

1.4 Discuss the function of each of the four types of memory (executive, scratch pad, control program, and data table) found in programmable controllers.

1.5 Explain the difference between volatile and nonvolatile memory.

1.6 Explain the purpose and function of the registers in a microprocessor.

1.7 Discuss the function and purpose of the various input/output modules used in programmable controller systems.

1.8 List some common applications for personal computers in programmable controller systems.

1.9 Discuss the various types of programming terminals and devices used in programmable controller systems.

1.10 Discuss the basic instructions used in ladder logic programs.

BIBLIOGRAPHY

1. Jones, C. T., and Bryan, L. A., *Programmable Controllers Concepts and Applications*, International Programmable Controls, Inc., First Edition, 1983.

2. *Modicon 584 Programmable Controllers User's Manual*, Gould Modicon Division, January 1982.

3. *PLC-2/30 Programmable Controller: Programming and Operations Manual.* Allen-Bradley, Publication 1772-6.8.3, 1984.

4. Hughes, T. A., *Measurement and Control Basics*, Instrument Society of America, 1988.

5. Gilbert, R. A., and Llewellyn, J. A., *Programmable Controllers—Practices and Concepts*, Industrial Training Corporation, 1985.

6. Editors of Time-Life Books, *Computer Basics*, Time-Life Books, 1985.

7. Plato Computer-Based Training, *Programmable Controller Fundamentals*, Allen-Bradley Company, 1985.

2

Numbering Systems and Codes

Introduction The application of programmable controllers to process control requires an understanding of numbering systems and codes. Programming, intra-system communications, and input/output (I/O) interfacing in particular require a solid background in the binary numbering system and codes.

Numbering Systems Many methods are used for encoding numbers in programmable controllers. The most commonly used in programmable controllers is the binary system, but the octal and hexadecimal numbering systems are also encountered in these applications. We will start with a brief review of the decimal numbering system, followed by coverage of the other three systems.

Decimal Numbering System The decimal numbering system is in common use probably because man started to count with his fingers. However, the decimal system is not easy to implement electronically. A ten-state electronic device would be quite costly and complex. It is much easier and more efficient to use the binary (two-state) numbering system when manipulating numbers using logic circuits.

A decimal number, N_{10}, can be written mathematically as:

$$N_{10} = d_n R^n + \ldots d_2 R^2 + d_1 R^1 + d_0 R^0 \tag{2-1}$$

where R is equal to the number of digit symbols used in the system. R is called the radix and is equal to 10 in the decimal system. The subscript 10 on the number N in Equation 2-1 indicates that it is a decimal number. However, it is common practice to omit this subscript in writing out decimal numbers. The decimal digits, $d_n \ldots d_2, d_1, d_0$, can assume the values of 0, 1, 2, 3, 4, 5, 6, 7, 8, or 9 in the decimal numbering system. For example, the decimal number 1735 can be written as:

$$1735 = 1 \times 10^3 + 7 \times 10^2 + 3 \times 10^1 + 5 \times 10^0$$

When written as 1735, the powers of ten are implied by positional notation. The value of the decimal number is computed by multiplying each digit by the weight of its position and summing the result. As we will see, this is true for all numbering systems; the decimal equivalent of any number can be calculated by multiplying the digit times its base raised to the power of the digit's position. The general equation is given below:

$$N_b = Z_n R^n + \ldots Z_2 R^2 + Z_1 R^1 + Z_0 R^0 \qquad (2\text{-}2)$$

where Z = the value of the digit
b = base of numbering system

Binary Numbering System   The binary numbering system uses the number 2 as the base, and the only allowable digits are 0 or 1. This is the basic numbering system for computers and programmable controllers, which are basically electronic devices that manipulate 0s and 1s to perform math and control functions. It was easier and more convenient to design digital computers that operate on two entities or numbers rather than the ten numbers used in the decimal world. Furthermore, most physical elements in the process environment have only two states, such as a pump on or off, a valve open or closed, a switch on or off, and so on.

A binary number follows the same format as a decimal one: the value of a digit is determined by where it stands in relation to the other digits in a number. In the decimal system, a 1 by itself is worth 1; placing it to the left of a zero makes the 1 worth 10, and putting it to the left of two zeros makes it worth 100. This simple rule is the foundation of the arithmetic. For example, numbers to be added or subtracted are first arranged so that their place columns line up.

In the decimal system, each position to the left of the decimal point indicates an increasing power of 10. In the binary system, each place to the left signifies an increased power of two, i.e., 2^0 is one, 2^1 is two, 2^2 is four, 2^3 is eight, and so on. So, finding the decimal equivalent of a binary number is simply a matter of noting which place columns the binary 1s occupy and adding up their values. A binary number also uses standard positional notation and is written as $b_n \ldots b_2, b_1, b_0$. The decimal equivalent of a binary number can be found using the equation:

$$N_{10} = b_n \times 2^n + \ldots b_2 \times 2^2 + b_1 \times 2^1 + b_0 \times 2^0 \qquad (2\text{-}3)$$

where R equals 2 in the binary system, and each binary digit (bit) can take on the value of 0 or 1.

The decimal equivalent of the binary number 10101 can be found as follows:

$$1 \times 2^4 + 0 \times 2^3 + 1 \times 2^2 + 0 \times 2^1 + 1 \times 2^0$$

or $(1 \times 16) + (0 \times 8) + (1 \times 4) + (0 \times 2) + (1 \times 1) = 21 \text{ (decimal)}$

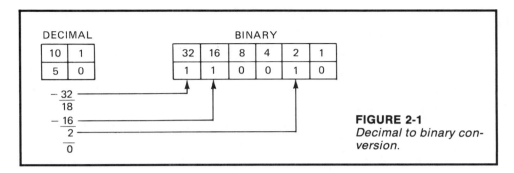

FIGURE 2-1
Decimal to binary conversion.

EXAMPLE 2-1

Problem: Convert the binary number 101110_2 to its decimal equivalent.

Solution: Using Equation 2-3:

$$N = b_n \times 2^n + \ldots b_2 \times 2^2 + b_1 \times 2^1 + b_0 \times 2^0$$

or

$$1 \times 2^5 + 0 \times 2^4 + 1 \times 2^3 + 1 \times 2^2 + 1 \times 2^1 + 0 \times 2^0 =$$
$$(1 \times 32) + (0 \times 16) + (1 \times 8) + (1 \times 4) + (1 \times 2) + (0 \times 1) =$$
$$32 + 0 + 8 + 4 + 2 + 0 = 46$$

To convert a decimal number to a binary number, we first subtract the largest possible power of two and continue subtracting the next largest possible power from the remainder, placing 1s in each column where this is possible and 0s where it is not. For example, we can convert the decimal number 50 to a binary number as shown in Figure 2-1.

EXAMPLE 2-2

Problem: Convert the decimal number 59 to a binary number.

Solution:

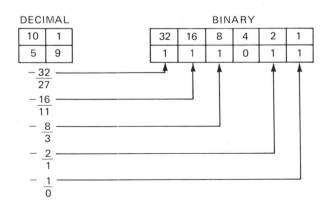

TABLE 2-1
Binary Addition

$0 + 0 = 0$
$0 + 1 = 1$
$1 + 0 = 1$
$1 + 1 = 10$

CARRY ⤴

The rules of addition for binary numbers are illustrated in Table 2-1. As in decimal addition, 0 plus 0 is 0 in binary. Similarly, 0 plus 1 equals 1, as it does in the decimal system. However, the third possibility of binary addition is different from decimal in that 1 plus 1 equals 0, with a carry of 1. In fact, the addition of numbers is performed in accordance with certain rules, regardless of which base (radix) is used to represent the numbers. When adding two binary numbers, the following algorithm is used:

1. Align the numbers so that their place or positional values correspond.

2. Starting with the least significant digit, add the pair of digits in this column.

3. If the sum is less than or equal to 1, it is expressed as a single binary digit and there is no carry.

4. If the sum of the two digits is greater than 1, it is expressed as a two-digit binary number. The more significant digit, 1, is carried to the next positional column, and the less significant digit is placed under the current column.

5. The digits in the next column, including any carry, are added until all columns have been computed.

EXAMPLE 2-3

Problem: Perform binary addition of the following numbers: 1010 and 0110.

Solution:

$$
\begin{array}{r}
111 \leftarrow \text{CARRY} \\
1010 \\
+\,0110 \\
\hline
10000
\end{array}
$$

The rules of binary subtraction are given in Table 2-2.

The following example problem will help illustrate binary subtraction.

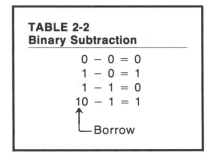

TABLE 2-2
Binary Subtraction

$$0 - 0 = 0$$
$$1 - 0 = 1$$
$$1 - 1 = 0$$
$$10 - 1 = 1$$

Borrow

EXAMPLE 2-4

Problem: Subtract the binary number 110_2 from the number 1001_2.

Solution:

$$\begin{array}{r} 1001 \\ -0110 \\ \hline 11 \end{array}$$

To this point, we have discussed only positive binary numbers. Several common methods are used to represent negative binary numbers in programmable controller systems. The first is *signed-magnitude binary*. This method places an extra bit (sign bit) in the leftmost position and lets this bit determine whether the number is positive or negative. The number is positive if the sign bit is 0 and negative if the sign bit is 1. For example, in a 16-bit machine, if we have a 12-bit binary number, $00000010101_2 = 21_{10}$, to express the positive and negative values, we would manipulate the leftmost or most significant bit. So, using the signed magnitude method: $000000000010101 = +21$ and $1000000000010101 = -21$.

Another common method used to express negative binary numbers is called *two's complement binary*. To complement a number means to change it to a negative number. For example, the binary number 10101 is equal to decimal 21. To get the negative using the two's complement method, you complement each bit and then add a 1 to the least significant bit.

In the case of the binary number $010101 = 21$, its two's complement would be: $101011 = -21$.

Octal Numbering System The binary numbering system requires substantially more digits to express a number than the decimal system. For example, $130_{10} = 10000010_2$, so it takes 8 binary digits to express a decimal number over 127. It is also difficult for people to read and manipulate large numbers without making errors. To reduce errors in binary number manipulations, some computer companies started using the octal numbering system. This system uses the number 8 as a base with the eight digits of 0, 1, 2, 3, 4, 5, 6,

7. Like all other number systems, each digit in an octal number has a weighted value according to its position. For example:

$$1301_8 = 1 \times 8^3 + 3 \times 8^2 + 0 \times 8^1 + 1 \times 8^0$$
$$= 1 \times 512 + 3 \times 64 + 0 \times 8 + 1 \times 1$$
$$= 512 + 192 + 0 + 1$$
$$= 705_{10}$$

The octal system is used as a convenient means of writing or manipulating binary numbers in computer systems. A binary number with a large number of 1s and 0s can be represented with an equivalent octal number with fewer digits. As shown in Table 2-3, one octal digit can be used to express three binary digits so that the number is reduced by a factor of three. For example, the following binary number can be represented as octal by grouping binary bits in groups of three starting with the least significant bit.

1	1	0	0	0	1	1	0	1	0	1	0	1	0	1	0	3-BIT GROUPS

1	4	3	2	5	2	OCTAL NUMBER

EXAMPLE 2-5

Problem: Represent the binary number 101011001101111 in octal.

Solution: To convert from binary to octal, we simply divide the binary number into groups of three bits, starting with the least significant bit (LSB) and using Table 2-3 to convert the 3-bit groups to their octal equivalent.

TABLE 2-3
Binary and Octal
Equivalent Numbers

Binary	Octal
000	0
001	1
010	2
011	3
100	4
101	5
110	6
111	7

1	0	1	0	1	1	0	0	1	1	0	1	1	1	1	3-BIT GROUPS
5			3			1			5			7			OCTAL NUMBER

Therefore, $101011001101111_2 = 53157_8$

EXAMPLE 2-6

Problem: Convert the decimal number 370_{10} to an octal number.

Solution: Decimal to octal conversion is obtained by successive division by the octal base number 8 as follows:

$$\frac{370}{8} = 46 + 2$$

$$\frac{46}{8} = 5 + 6$$

$$\frac{5}{8} = 0 + 5$$

$$370_{10} = 562_8$$

Hexadecimal Numbering System   The hexadecimal numbering system provides an even shorter notation than the octal system and is the most commonly used numbering system in computer applications. The hexadecimal system has a base of 16, and four binary bits are used to represent a single symbol. The sixteen symbols are 0, 1, 2, 3, 4, 5, 6, 7, 8, 9, A, B, C, D, E, and F. The letters A through F are used to represent the binary numbers 1010, 1011, 1100, 1101, 1110, and 1111, corresponding to the decimal numbers 10 through 15. The hexadecimal digits and their binary equivalents are given in Table 2-4.

TABLE 2-4
Hexadecimal and Binary Equivalents Numbers

Hexadecimal	Binary	Hexadecimal	Binary
0	0000	9	1001
1	0001	A	1010
2	0010	B	1011
3	0011	C	1100
4	0100	D	1101
5	0101	E	1110
6	0110	F	1111
7	0111	10	10000
8	1000	11	10001

To convert a binary number to a hexadecimal number, we use Table 2-4. For example, the binary number 0110 1111 1000 is simply the hex number 6F8.

Again, the hexadecimal numbers follow the standard positional convention $H_n \ldots H_2, H_1, H_0$, where the positional weights for hexadecimal numbers are powers of sixteen with 1, 16, 256, and 4096 being the first four decimal values.

To convert from hexadecimal to decimal numbers, we use the formula:

$$N_{10} = Z_n 16^n + \ldots Z_2 16^2 + Z_1 16^1 + Z_0 16^0$$

EXAMPLE 2-7

Problem: Convert the hex number 1FA to its decimal equivalent.

Solution: $N_{10} = Z_n 16^n + \ldots Z_2 16^2 + Z_1 16^1 + Z_0 16^0$

Since $Z_2 = 1$, $Z_1 = F_H = 15_{10}$, and $Z_0 = A_H = 10_{10}$,

we obtain

$$1 \times 16^2 + 15 \times 16^1 + 10 \times 16^0 = 256 + 240 + 10 = 506$$

To convert a decimal number to a hexadecimal number, we use the following algorithm:

1. Divide the decimal number by 16 and record the quotient and the remainder.

2. Divide the quotient from the division in step 1 by 16 and record the quotient and the remainder.

3. Repeat step 2 until the quotient is zero.

4. The hexadecimal equivalents of the remainders generated by the divisions are the digits of the hexadecimal number, where the first remainder is the least significant digit (LSD) and the last remainder is the most significant digit (MSD). An example will illustrate the conversion from decimal to hexadecimal numbers.

EXAMPLE 2-8

Problem: Convert the decimal number 610 to a hexadecimal number.

Solution:

Division	Quotient	Remainder
610/16	38	$2_{10} = 2_{16}$ (LSD)
38/16	2	$6_{10} = 6_{16}$
2/16	0	$2_{10} = 2_{16}$

Therefore, $610_{10} = 262_{16}$.

We can check the answer by converting 262_{16} back to decimal as follows:

$$N_{10} = 2 \times 16^2 + 6 \times 16^1 + 2 \times 16^0$$
$$= 2 \times 256 + 6 \times 16 + 2 \times 1$$
$$= 512 + 96 + 2 = 610$$

Hexadecimal Arithmetic

The addition of numbers is performed according to certain rules, regardless of which base (radix) is used to represent the numbers. When adding two hexadecimal numbers, the following algorithm is used:

1. Align the numbers so that their place positional values correspond.

2. Starting with the least significant digit, add the pair of digits in this column.

3. If the sum is less than sixteen decimal, it is expressed as a single hexadecimal digit and there is no carry.

4. If the sum of the two digits is greater than or equal to decimal sixteen, it is expressed as a two-digit hexadecimal number. The more significant digit, 1, is carried to the next positional column, and the less significant digit is placed under the current column.

5. The digits in the next column, including any carry, are added until all columns have been computed.

EXAMPLE 2-9

Problem: Add the hexadecimal numbers 1736_{16} and $1AC_{16}$.

Solution: Align the numbers so that the positional values correspond.

$$\begin{array}{r} 1736 \\ +1AC \\ \hline \end{array}$$

Starting with the rightmost (16^0) positional value, add the pair of digits in this column.

$$\begin{array}{r} 1 \leftarrow \text{Carry} \\ 1736 \\ +1AC \\ \hline 8 \end{array}$$

Since the sum of C + 6 is more than 15_{10}, there is a carry to the next column.

Continue with the next column. Here 1 + 3 + A = E, so there is no carry to the last column.

$$\begin{array}{r} 1 \\ 1736 \\ +1AC \\ \hline E\,8 \end{array}$$

Continue in the same manner, adding the last two columns.

$$1736$$
$$+1AC$$
$$18E8$$

The subtraction of numbers is performed using the same process for all numbering systems. It is basically a reversal of the addition process. The most convenient method for subtracting a hexadecimal number is to convert each digit above 9 to decimal and use standard decimal subtraction. For example, confronted with the problem $F - A$, convert to $15 - 10$ to arrive at the solution, as follows:

$$F \quad (15_{10})$$
$$\underline{-A} \quad (10_{10})$$
$$5$$

Borrowing is performed in the same manner as in the decimal numbering system. But we must remember that a borrow of 1 in the hexadecimal system is equivalent to 16 in the decimal system.

EXAMPLE 2-10

Problem: Subtract 29_{16} from 68_{16}.

Solution: To solve the problem of $68_{16} - 29_{16}$, we first note that 8 is less than 9, so we must borrow a 1 from the 16^1 column. Solving the problem in decimal, we have

$$\text{the value borrowed} \rightarrow 16_{10} + 8_{10} - 9_{10}$$
$$= 24_{10} - 9_{10}$$
$$= 15_{10}$$
$$= F_{16}$$
$$\text{borrow} \rightarrow 1$$
$$68$$
$$\underline{-29}$$
$$F$$

Next, we have $5 - 2 = 3$ in the left column, where the 6 is reduced to 5 since we borrowed from this column.

$$5$$
$$68$$
$$\underline{-29}$$
$$3F$$

It is important to note that the hexadecimal system is used for human convenience only, and the computer system actually converts the hex numbers into binary strings and operates on the binary digits.

Data Codes
Data codes translate information (alpha, numeric, or control characters) to a form that can be transferred electronically and then converted back to its original form. A code's efficiency is a measure of its ability to utilize the maximum capacity of the bits and to recover from error. In the evolution of the various codes, there has been a steady increase in their efficiency to transfer data. A brief discussion of the commonly used codes follows.

Binary Code   It is possible to represent 2^n different symbols in a purely binary code of n bits. The binary code is a direct conversion of the decimal number to the binary. This is illustrated in Table 2-5. It is the most commonly used code in computers because it is a systematic arrangement of the digits. It is also a weighted code, where each column has a magnitude of 2^n associated with it, and it is easy to translate. In Table 2-5 note that the least significant bit (LSB) alternates every time, whereas the second least significant bit repeats every two times, the third least significant bit repeats every four times, and so on.

Baudot Code   The Baudot code was the first successful data communications code. It uses five consecutive bits and an additional 2½ start/stop bits to represent a data character. It transfers asynchronous data at a very slow rate (10 characters per second) using punched paper tape. A disadvantage of Baudot code is that it can represent only the 58 characters shown in Table 2-6.

Other limitations of Baudot code are the sequential nature of the code, the high overhead, and the lack of error detection.

BCD Code   As computer and data communications technology improved, more efficient codes were developed, specifically the synchronous binary coded decimal series.

**TABLE 2-5
Binary Code**

Decimal	Binary	Decimal	Binary
0	0000	12	1100
1	0001	13	1101
2	0010	14	1110
3	0011	15	1111
4	0100	16	10000
5	0101	17	10001
6	0110	18	10010
7	0111	19	10011
8	1000	20	10100
9	1001	21	10101
10	1010	22	10110
11	1011	23	10111

TABLE 2-6
Baudot Code

Character Case		Bit Pattern	Character Case		Bit Pattern
Lower	Upper	54321	Lower	Upper	54321
A	-	00011	Q	1	10111
B	?	11001	R	4	01010
C	:	01110	S	'	00101
D	$	01001	T	5	10000
E	3	00001	U	7	00111
F	!	01101	V	;	11110
G	&	11010	W	2	10011
H	#	10100	X	/	11101
I	8	00110	Y	6	10101
J	Bell	01011	Z	"	10001
K	(	01111	Letters	(Shift) ↓	11111
L	)	10010	Figures	(Shift) ↑	11011
M	.	11100	Space	(Sp) =	00100
N	,	01100	Carriage Return <		01000
O	9	11000	Line Feed ___		00010
P	0	100110	Blank		00000

This BCD series consists of the following: (1) standard binary coded decimal (BCD), (2) binary coded decimal interchange code (BCDIC), (3) extended binary coded decimal (EBCD), and (4) extended binary coded decimal interchange code (EBCDIC).

The BCD code was first used to perform internal numeric calculations within data processing devices. The BCD code is commonly used in programmable controllers to code data to numeric light-emitting diode (LED) displays and from panel-mounted digital thumb wheel units. Its main disadvantages are that it has no alpha characters and no error checking capability. A listing of the BCD code for decimal numbers from 0 to 10 is given in Table 2-7.

EXAMPLE 2-11

Problem: Convert the following decimal numbers to BCD: (a) 276; (b) 567; (c) 719; and (d) 4500

Solution: Using Table 2-7, the decimal numbers can be expressed in BCD code as follows:
(a) 276 = 0010 0111 0110
(b) 567 = 0101 0110 0111
(c) 719 = 0111 0001 1000
(d) 4500 = 0100 0101 0000 0000

TABLE 2-7
BCD Code

Numeric (Decimal)	Binary Equivalent	BCD Code
0	0	0000
1	1	0001
2	10	0010
3	11	0011
4	100	0100
5	101	0101
6	110	0110
7	111	0111
8	1000	1000
9	1001	1001
10	1010	0001 0000

BCDIC Code The binary coded decimal interchange code was developed to allow for both alpha and numeric transfer of data between systems. This 6-bit binary code has 64 alphanumeric characters but no provision for error checking (see Table 2-8).

EXAMPLE 2-12

Problem: Express the message "ΔP is 100 in." in BCDIC Code.

Solution: Using Table 2-8, find the binary code as follows:

Δ = 101111, P = 100111, Space = 000000, I = 111001, S = 010010, Space = 000000, 1 = 000001, 0 = 001010, 0 = 001010, Space = 000000, I = 111001, N = 100101, . = 001100

EBCD Code A slightly different version of BCD, extended binary coded decimal was first developed for the IBM electric typewriter. Information was transferred to standard paper, and, at the same time, a EBCD code was sent over a communications line via punched paper tape. This code has 6 bits for data and a parity bit for sensing transmission errors. This sequential binary code uses uppercase and lowercase as shown in Table 2-9.

This code contains the following non-printing characters:

BS	—Backspace	HT	—Horizontal tab
B	—Bypass	IL	—Idle
DEL	—Delete	LC	—Lowercase
EOB	—End of block	LF	—Line feed
EOT	—End of transmission	NL	—New line

TABLE 2-8
BCDIC Code

Bits	3 2 1	0 0 0	0 0 1	0 1 0	0 1 1	1 0 0	1 0 1	1 1 0	1 1 1
6 5 4									
0 0 0		SP	1	2	3	4	5	6	7
0 0 1		8	9	0	# =	@ .	:	>	√
0 1 0		SP	/	S	T	U	V	W	X
0 1 1		Y	Z	= = =	,	% (	γ	\	⧻
1 0 0		-	J	K	L	M	N	O	P
1 0 1		Q	R	!	$	*	]	;	△
1 1 0		& +	A	B	C	D	E	F	G
1 1 1		H	I	?	●	□)	[	<	≢

NOTE: COMMERCIAL USAGE → | &
+ | ← SCIENTIFIC USAGE

P	—Parity bit	RES	—Restore
PF	—Punch off	RS	—Reader stop
PN	—Punch on	SP	—Space
PRE	—Prefix	UC	—Uppercase

EBCDIC Code   The extended binary coded decimal interchange code seemed a step backward in the development of data communications codes, because it had no parity bit for error checking and not all the defined characters were required. However, external error checking was employed in most systems that used the code. Its main advantage was that its 8-bit code accommodated 256 combinations (see Table 2-10).

The mnemonics in Table 2-10 have the following meanings:

BS	= Backspace	HT	= Horizontal tab
BYP	= Bypass	IL	= Idle
DEL	= Delete	LC	= Lowercase
EOB	= End of block	LF	= Line feed
EOT	= End of transmission	PF	= Punch off

TABLE 2-9
Extended BCD Code

Bits	6 5 4	0 0 0	0 0 1	0 1 0	0 1 1	1 0 0	1 0 1	1 1 0	1 1 1
3 2 1									
0 0 0		SP / SP	: / 4	/ 2	, / 6	= / 1	% / 5	; / 3	/ 7
0 0 1		— / -	M / m	K / k	O / o	J / j	N / n	L / l	P / p
0 1 0		¢ / @	U / u	S / s	W / w	? / /	V / v	T / t	X / x
0 1 1		+ / &	D / d	B / b	F / f	A / a	E / e	C / c	G / g
1 0 0		. / 8	PN / PN	) / φ	UC / UC	(/ 9	RS / RS	" / #	EOT / EOT
1 0 1		Q / q	RES / RES		BS / BS	R / r	NL / NL	! / $	IL / IL
1 1 0		Y / y	BY / BY	EOB / EOB		Z / z	LF / LF	{ / ,	PRE / PRE
1 1 1		H / h	PF / PF		LC / LC	I / i	HT / HT	⌐ / .	DLE / DLE

NOTE: UPPER CASE → [ ] ← LOWER CASE

PN = Punch on   SM = Start message
PRE = Prefix (ESC)   SP = Space
RES = Restore   UC = Uppercase
RS = Reader stop   NL = New line

ASCII Code   The most widely used code is ASCII, the American Standard Code for Information Interchange, which was developed in 1963. This code has 7 bits for data (allowing 128 characters) (see Table 2-11). The ASCII code can operate synchronously or asynchronously with 1 or 2 stop bits. ASCII format has 32 control characters (see Table 2-12). These control codes are used to indicate, modify, or stop a control function in the transmitter or receiver. Seven of the ASCII control codes are called *format effectors*. They pertain to the control of a printing device. The use of format effectors increases code efficiency and speed by replacing frequently used character combinations with a single code. The following is a list of the format effectors used in the ASCII code: BS

TABLE 2-10
EBCDIC Code

Bits 4321 \ 8765	0000	0001	0010	0011	0100	0101	0110	0111	1000	1001	1010	1011	1100	1101	1110	1111
0000	NUL	DLE	DS		SP	&	-				SMM	VT	FF	CR	SO	SI
0001	SCH	DC1	SOS				/		a	j	CC		IFS	IGS	IRS	IUS
0010	STX	DC2	FS	SYN					b	k	SM		DC	ENQ	ACK	BEL
0011	ETX	DC3							c	l	t			NAK		SUB
0100	PF	RES	BYP	PN					d	m	u	.	<	(	+	|
0101	HT	NL	LF	RS					e	n	v	$	*	)	;	¬
0110	LC	BS	EOB	UC					f	o	w	,	%	_	>	?
0111	DEL	IL	PRE	EOT					g	p	x	#	@	'	=	"
1000	CAN								h	q	y		H	Q	Y	8
1001	EM								i	r	z		I	R	Z	9
1010					¢	!	|	:								
1011																
1100									A	J			C	L	T	3
1101									B	K	S		D	M	U	4
1110													E	N	V	5
1111	0	1	2	3	4	5	6	7	8	9			F	O	W	□

TABLE 2-11
ASCII Code

Bits	7 6 5		0 0 0	0 0 1	0 1 0	0 1 1	1 0 0	1 0 1	1 1 0	1 1 1
4 3 2 1										
0 0 0 0			NUL	DLE	SP	0	@	P	`	p
0 0 0 1			SOH	DC1	!	1	A	Q	a	q
0 0 1 0			STX	DC2	"	2	B	R	b	r
0 0 1 1			ETX	DC3	#	3	C	S	c	s
0 1 0 0			EOT	DC4	$	4	D	T	d	t
0 1 0 1			ENQ	NAK	%	5	E	U	e	u
0 1 1 0			ACK	SYN	&	6	F	V	f	v
0 1 1 1			BEL	ETB	'	7	G	W	g	w
1 0 0 0			BS	CAN	(	8	H	X	h	x
1 0 0 1			HT	EM	)	9	I	Y	i	y
1 0 1 0			LF	SUB	*	:	J	Z	j	z
1 0 1 1			VT	ESC	+	;	K	[	k	{
1 1 0 0			FF	FS	,	<	L	\	l	|
1 1 0 1			CR	GS	-	=	M	]	m	}
1 1 1 0			SO	RS	.	>	N	^	n	~
1 1 1 1			SI	US	/	?	O	—	o	DEL

TABLE 2-12
Legend for ASCII Control Characters

Symbol	Name	Symbol	Name
NUL	NULL	DLE	Data link escape
SOH	Start of heading	DC1	Device control 1
STX	Start of text	DC2	Device control 2
ETX	End of text	DC3	Device control 3
EOT	End of transmission	DC4	Device control 4
ENQ	Inquiry	NAK	Negative acknowledge
ACK	Acknowledge	SYN	Synchronous idle
BEL	Bell	ETB	End trans. block
BS	Backspace	CAN	Cancel
HT	Horizontal tabulation	EM	End of medium
LF	Line feed	SUB	Substitute
VT	Vertical tabulation	ESC	Escape
FF	Form feed	FS	File separator
CR	Carriage return	GS	Group separator
SO	Shift out	RS	Record separator
SI	Shift in	US	Unit separator
DEI	Delete	SP	Space

(Backspace), HT (Horizontal tab), LF (Line feed), VT (Vertical tab), FF (Form feed), CR (Carriage return), and SP (Space).

An example problem will aid in the understanding of the ASCII code.

EXAMPLE 2-13

Problem: Express the words PUMP ON using ASCII code.

Solution: Using Table 2-11, we obtain the binary code for each character as follows:

Bits 7 6 5 4 3 2 1	Character
1 0 1 0 0 0 0	P
1 0 1 0 1 0 1	U
1 0 0 1 1 0 1	M
1 0 1 0 0 0 0	P
0 1 0 0 0 0 0	(space)
0 1 1 0 0 0 1	1

The ASCII code is so widely used that most programmable controller manufacturers offer an ASCII input module with their systems.

EXERCISES

2.1 Convert the binary number 10010 to a decimal number.

2.2 Find the binary equivalent to the decimal number 115.

2.3 Add the binary number 10101 to the binary number 1101 and find the decimal result.

2.4 Convert the following decimal numbers to their hexadecimal equivalents: (a) 656_{10}, (b) 2566_{10}, (c) 270_{10}, (d) 29_{10}, and (e) $12,056_{10}$.

2.5 Convert the following hexadecimal numbers to their decimal equivalents: (a) 57_{16}, (b) 754_{16}, (c) 345_{16}, (d) 5687_{16}, and (e) 455_{16}.

2.6 Convert the following binary numbers to their hexadecimal equivalents: (a) 101_2, (b) 1011111_2, (c) 1011100_2, (d) 01101_2 and (e) 11001111101_2.

2.7 Add the following hexadecimal numbers: (a) A34 + E45, (b) C87 + AA1, (c) 34 + CCE, (d) 397 + 1A2, and (e) 3D4 + 167.

2.8 Perform the following subtractions on hexadecimal numbers: (a) B8 − 25, (b) 36 − A, (c) 1C1 − FC, (d) 300 − 3, and (e) 2BE − 234.

BIBLIOGRAPHY

1. Kintner, P. M., *Electronic Digital Techniques*, McGraw-Hill Book Company, 1968.

2. Floyd, T. L., *Digital Logic Fundamentals*, Charles E. Merrill Publishing Company, 1977.

3. Malvino, A. P., *Digital Computer Electronics—An Introduction to Microcomputers*, Second Edition, McGraw-Hill Book Company, 1983.

4. Wickers, W. E., *Logic Design with Integrated Circuits*, John Wiley & Sons, Inc., 1968.

3

Logic System Fundamentals

Introduction   A knowledge of digital logic principles is required in the design and implementation of logic systems and microprocessor-based control systems. In this chapter, the basic concepts of logic gates, digital devices, and logic design will be discussed.

Logic Gates   In most logic systems, binary numbers 1 and 0 are represented by voltage or current levels. For example, in transistor-transistor logic (TTL) gates, a binary 1 is represented by a voltage signal in the range of 2.0 to 5.0 volts, and a binary 0 is represented by a voltage level between 0 and 0.8 volt. Solid-state electronic circuits are available that can be used to manipulate digital signals to perform a variety of logical functions, such as AND, OR, NAND, NOR, NOT, and exclusive OR. These circuits are the building blocks used in logic system design.

In subsequent sections of this chapter, these logic circuits or gates are defined and a symbol is given for each. We will also cover some typical applications of logic gates such as parity generators and binary adder circuits.

OR Gates   An OR gate, with two or more inputs and a single output, operates in accordance with the following definition: *The output of an OR gate assumes the 1 state if one or more inputs assume the 1 state.*

The inputs to a logic gate are designated by A, B, . . . , N and the output by Z. It is assumed that the inputs and outputs can take one of two possible values, either 0 or 1. A standard symbol for the OR gate is given in Figure 3-1 together with the logic expression for this gate (i.e., $Z = A + B + \ldots + N$). A truth table, which contains a tabulation of all possible input values and their corresponding outputs, is also given for a two-input OR gate.

The following is an example of the use of OR logic in process control: If the water level in a hot water heater is low, OR the temperature in the tank is too high, a logic system can be designed to turn off the heater in the system.

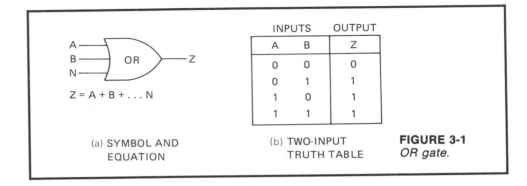

| INPUTS | | OUTPUT |
A	B	Z
0	0	0
0	1	1
1	0	1
1	1	1

$Z = A + B + \ldots N$

(a) SYMBOL AND EQUATION

(b) TWO-INPUT TRUTH TABLE

FIGURE 3-1
OR gate.

The following logic identities for OR gates can be easily verified.

$$A + B + C = (A + B) + C = A + (B + C) \qquad (3\text{-}1)$$

$$A + B = B + A \qquad (3\text{-}2)$$

$$A + A = A \qquad (3\text{-}3)$$

$$A + 1 = 1 \qquad (3\text{-}4)$$

$$A + 0 = A \qquad (3\text{-}5)$$

Remember that A, B, and C can take on only the value of 0 or 1. Refer to the definition of the OR gate and to the truth table in Figure 3-1.

EXAMPLE 3-1

Problem: Implement the logic function $A + B + C$ using two-input OR Gates.

Solution: The implementation of $A + B + C$ is shown in Figure 3-2.

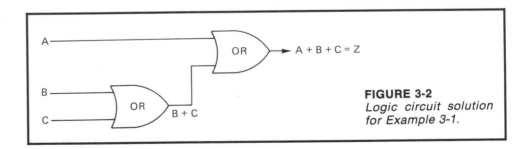

FIGURE 3-2
Logic circuit solution for Example 3-1.

NOT Gate   The NOT or inverter gate produces an output opposite the input. Figure 3-3 shows the operation of the NOT gate. This logic gate is used to invert or complement a logic function.

A typical logic circuit application would be if we had a temperature switch

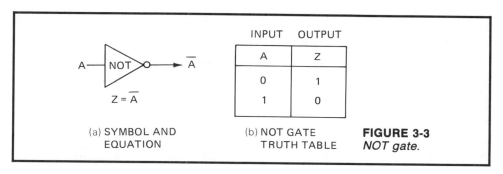

(a) SYMBOL AND
EQUATION

(b) NOT GATE
TRUTH TABLE

FIGURE 3-3
NOT gate.

low (TSL) on a process heater but need the NOT low temperature ($\overline{\text{TSL}}$) to perform a process control operation. In that case we would use a NOT gate to invert the TSL signal.

AND Gate   An AND gate has two or more inputs and a single output, and it operates in accordance with the following definition: *The output of an AND gate assumes the 1 state if and only if all the inputs assume the 1 state.*

A standard symbol for the AND gate is given in Figure 3-4 together with the logic expression and a two-input AND gate truth table.

The logic expressions for an AND gate are as follows:

$$ABC = (AB)C = A(BC) \tag{3-6}$$

$$AB = BA \tag{3-7}$$

$$AA = A \tag{3-8}$$

$$A1 = A \tag{3-9}$$

$$A0 = 0 \tag{3-10}$$

These identities can be verified by reference to the definitions of the AND gate and by using a truth table for the AND Gate.

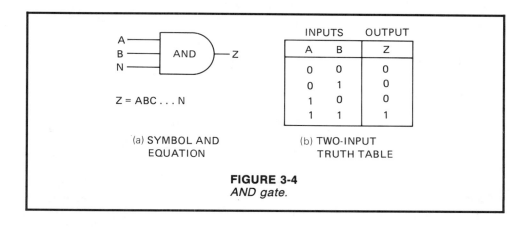

(a) SYMBOL AND
EQUATION

(b) TWO-INPUT
TRUTH TABLE

FIGURE 3-4
AND gate.

EXAMPLE 3-2

Problem: Verify the identity A1 = A.

Solution: A1 = A can be verified as follows: let B = 1 and tabulate AB = Z in a truth table.

A	B	Z
0	1	0
1	1	1

Note that A = Z, so that A1 = A for both values (0 and 1) of A.

Some important auxiliary identities used in logic design are as follows:

$$A + AB = A \qquad\qquad (3\text{-}11)$$

$$A + \overline{A}B = A + B \qquad\qquad (3\text{-}12)$$

$$(A + B)(A + C) = A + BC \qquad\qquad (3\text{-}13)$$

EXAMPLE 3-3

Problem: Verify the identity A + AB = A.

Solution: A + AB = A can be verified as follows: First tabulate AB = Z in a truth table.

A	B	Z
0	0	0
0	1	0
1	0	0
1	1	1

Then tabulate A + Z in a truth table.

A	Z	Y
0	0	0
0	0	0
1	0	1
1	1	1

Note that A = Y, so that A + AB = A.

These identities are important because they reduce the number of logic gates required to synthesize a logic function. For example, in Equation 3-13, the left side (A + B)(A + C) can be synthesized using the three logic gates shown in Figure 3-5. However, if the reduced form on the right side of the equation is used (A + BC), only two gates are required to produce the output Z, as shown in Figure 3-6.

NAND Gate   Another logic gate of interest is the NAND gate; its operations are summarized in Figure 3-7. Note that the NAND gate output is the exact

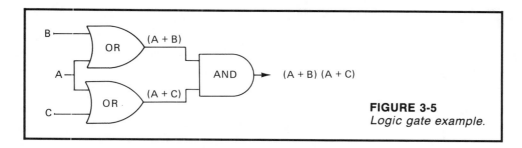

FIGURE 3-5
Logic gate example.

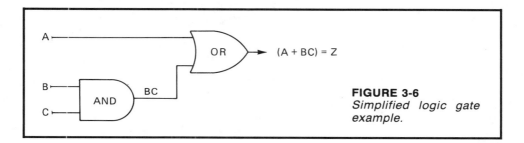

FIGURE 3-6
Simplified logic gate example.

opposite of the AND gate output. When all the inputs to a NAND are 1, the output is 0. In all other configurations, the NAND output has a logic value of 1.

NOR Gate   Another logic gate in common use is the NOR gate. Figure 3-8 shows its operations. It produces a logic 1 result if and only if all inputs are logic 0. The truth table shown is for a two-input NOR gate.

Exclusive OR Gate   Another helpful logic device is the two-input Exclusive OR gate. This gate provides a logic 1 output if and only if the input signals are not identical. The function of the Exclusive OR (XOR) is illustrated in

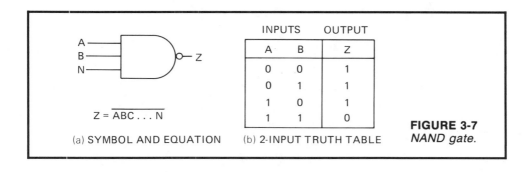

FIGURE 3-7
NAND gate.

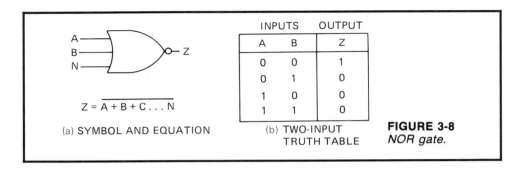

(a) SYMBOL AND EQUATION

(b) TWO-INPUT TRUTH TABLE

$Z = \overline{A + B + C \ldots N}$

INPUTS		OUTPUT
A	B	Z
0	0	1
0	1	0
1	0	0
1	1	0

FIGURE 3-8
NOR gate.

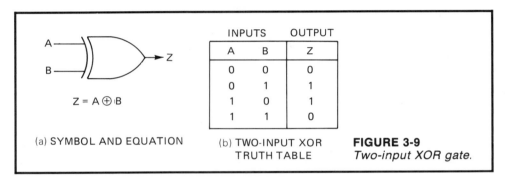

(a) SYMBOL AND EQUATION

(b) TWO-INPUT XOR TRUTH TABLE

$Z = A \oplus B$

INPUTS		OUTPUT
A	B	Z
0	0	0
0	1	1
1	0	1
1	1	0

FIGURE 3-9
Two-input XOR gate.

Figure 3-9. The logic operator symbol ⊕ is normally used for the Exclusive OR operation.

Logic Laws and Identities

A great deal of logic design is based upon a set of rules called DeMorgan's Theorem. This theorem demonstrates that any logic equation can be synthesized from NAND gates or from NOR gates. DeMorgan's theorem can be summarized: *If the inversion bar (NOT function) is broken between two logic variables, the logic symbol connecting the variable can be changed.*

The reverse is also true: *If the inversion bar is joined between two variables, the logic symbol connecting the variables can be changed.*

In equation form, DeMorgan's Theorem is:

$$\overline{(A + B + \ldots N)} = \overline{A}\,\overline{B} \ldots \overline{N} \qquad (3\text{-}16)$$

$$\overline{(AB \ldots N)} = \overline{A} + \overline{B} \ldots \overline{N} \qquad (3\text{-}17)$$

The practical importance of this law is that a single NOT function can be used to convert an AND function to an OR function or the inverse, so that NOR or NAND gates can be used in a single logic application. This is normally more cost-effective, since logic gates are normally furnished in integrated circuits with four or more gates in a single package.

TABLE 3-1
Logic Laws, Theorems and Identities

Fundamental Laws

OR	AND	NOT
$A + 0 = A$	$A0 = 0$	$A + \overline{A} = 1$
$A + 1 = 1$	$A1 = A$	$A\overline{A} = 0$
$A + A = A$	$AA = A$	$\overline{\overline{A}} = A$ (involution)

Associative Laws

$$(A + B) + C = A + (B + C), (AB)C = A(BC)$$

Commutative Laws

$$A + B = B + A, AB = BA$$

Distributive Laws

$$A(B + C) = AB + AC$$

DeMorgan's Theorem

$$\overline{(A + B + \ldots N)} = \overline{A}\,\overline{B}\ldots, \overline{N} \text{ and } \overline{AB\,N} = \overline{A} + \overline{B} + \overline{N}$$

Auxiliary Identities

$$A + AB = A \text{ and } A + \overline{A}B = A + B$$
$$(A + B)(A + C) = A + BC$$

DeMorgan's Theorem and some other basic logic laws and identities have been summarized in Table 3-1 for easy reference.

Three laws are very useful in logic design: distributive, involution ($\overline{\overline{A}} = A$), and DeMorgan's Theorem. Distribution is essentially equivalent to factoring. Involution and DeMorgan's Theorem work together to interchange AND and OR functions.

In applying these laws to logic design, the least complex gate realization is found by factoring out all common terms. The application of these rules to logic circuit design can be demonstrated by the following examples.

EXAMPLE 3-4

Problem: Simplify the logic function $Z = DC + ABC + B\overline{C}A + \overline{C}D$ using logic laws, theorems, and/or identities.

Solution:

$$Z = DC + ABC + B\overline{C}A + \overline{C}D$$

First we can use the commutative laws to rearrange the logic function as follows:

$$Z = CD + \overline{C}D + ABC + AB\overline{C}$$

Then use the associative laws to obtain $Z = (C + \overline{C})D + AB(C + \overline{C})$. Since $(C + \overline{C}) = 1$, the logic equation simplifies to $Z = D + AB$.

If we have an application where we only want to use a single type of logic gate such as a NAND gate or a NOR gate to synthesize a logic function, we can use DeMorgan's Theorem as shown in Example 3-5.

EXAMPLE 3-5

Problem: Synthesize $(A + B)(C + \overline{D})$ with NAND gates only.

Solution:

$$(A + B)(C + \overline{D}) = \overline{\overline{(A + B)(C + \overline{D})}} \text{ by involution}$$

$$= \overline{\overline{(\overline{A}\ \overline{B})(\overline{C}\ D)}} \text{ by using DeMorgan's Theorem twice.}$$

This result is synthesized using the NAND gates shown in Figure 3-10.

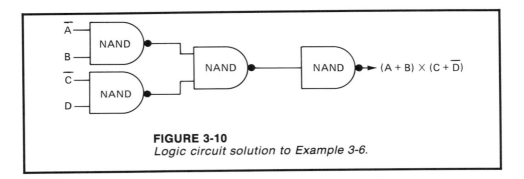

FIGURE 3-10
Logic circuit solution to Example 3-6.

Logic Circuits Design

We have discussed the synthesis of a logic equation by first reducing it using logic laws, theorems, and identities. In this section, we will discuss several common logic circuit design applications. The first application is the design of a parity bit generator.

Parity Bit Generator   A very practical logic circuit, the parity generator is used typically as a check on the transmission of binary data, such as over a telephone line. Parity generation consists of basing the logic output upon whether the number of 1s in the transmitted data is odd or even.

Assume a design is needed for an even four-input parity bit generator as shown in Figure 3-11. This circuit will produce a logic 1 output when there are an odd number of 1s in the input to the circuit. The truth table for this parity generator is given in Figure 3-12.

When the laws of logic are applied to this truth table, we obtain very complex logic expressions. However, we can look at the problem from another point of view by considering the problem with only two inputs. The truth table for an even parity bit generator with two inputs is given in Figure 3-13.

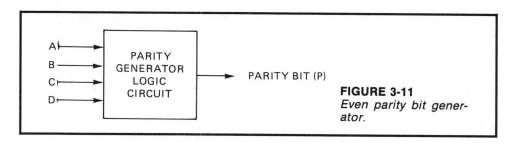

FIGURE 3-11
Even parity bit generator.

INPUTS				PARITY BIT	INPUTS				PARITY BIT
A	B	C	D	P	A	B	C	D	P
0	0	0	0	0	0	0	0	1	1
1	0	0	0	1	1	0	0	1	0
0	1	0	0	1	0	1	0	1	0
1	1	0	0	0	1	1	0	1	1
0	0	1	0	1	0	0	1	1	0
1	0	1	0	0	1	0	1	1	1
0	1	1	0	0	0	1	1	1	1
1	1	1	0	1	1	1	1	0	1

FIGURE 3-12
Truth table for parity bit generator.

Examination of this table reveals that it is the truth table for the Exclusive OR logic operation. Thus, the logic equation for a parity bit generator, if the total of the two input bits is odd, is:

$$P = A \oplus B$$

INPUTS		PARITY BIT
A	B	P
0	0	0
0	1	1
1	0	1
1	1	0

FIGURE 3-13
Two-input even parity generator truth table.

Now consider the result of the application of an additional Exclusive OR operation on P and a new input C.

$$
\begin{array}{lll}
\text{A} & 0101 & 0101 \\
\text{B} & 0011 & 0011 \\
\text{C} & \underline{0000} & \underline{1111} \\
\text{P} = \text{A} \oplus \text{B} = & 0110 & 0110 \\
\text{P} \oplus \text{C} = & 0110 & 1001
\end{array}
$$

It can be seen that a new parity bit is produced if there are an odd number of bits at the input. It may be concluded that an even parity generator can be based on the cascading of Exclusive OR operations. Therefore, a four-input even parity generator can easily be synthesized as shown in Figure 3-14 using only three logic gates. Note that the circuit is called an even parity generator because the parity bit is attached to the string of input bits to produce a data string with an even number of 1 bits.

Binary Adder   A practical application of logic gates is the operation of adding two binary numbers, A and B. First we will consider only the simple case where A and B each consists of one binary digit. The diagram of the logic circuit that will carry out the binary addition is shown in Figure 3-15 and the truth table is given in Figure 3-16. In the logic circuit, inputs A and B are each connected to an AND gate, A to gate G3, B to gate G4; they are also each connected to the opposite gate through NOT gates G1 and G2. Thus, when A is 1, its input to AND gate G3 is 1, and its input to AND gate G4 is 0; when A is 0, its input to G3 is 0, and to G4 is 1. The outputs from the two AND gates are connected to an OR gate, and the output from the OR gates is the binary sum of A and B. The two inputs are also connected into a third AND gate, G5, and the output from this gate produces the carry bit.

The four possible states of the half adder circuit are given in the truth table of Figure 3-16. In the first state A = 0 and B = 0, so none of the AND gates can give an output, since at least one input to each of them will be 0. Therefore, both the sum and carry outputs will be 0, giving the result 00.

In the second state, A = 0 and B = 1, so B gives a 1 into AND gate G4, and the 0 at input A is inverted by G2 to give another 1 into AND gate G4.

FIGURE 3-14
Four-input even parity generator.

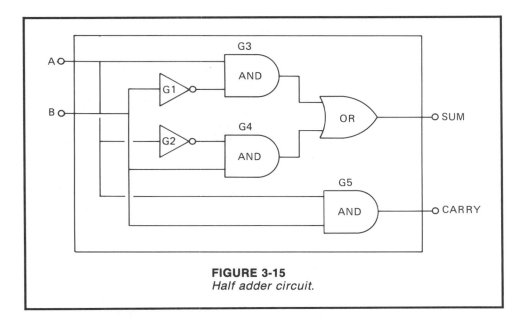

FIGURE 3-15
Half adder circuit.

Thus both inputs to G4 are 1, and the AND gate will produce a 1 output into the OR gate. Since both inputs to AND gate G3 are 0, the output will be 0. However, one of the inputs to the OR gate is 1, so the output will give a sum of 1. The carry AND gate has a 0 output, since one of the inputs is 0, giving the result of 01 from the half adder if A = 0 and B = 1.

In the third state (case A = 1 and B = 0), the operation is the same as for state 2, with inputs to G3 and G4 reversed, so the result is again 01.

In the final state, A = 1 and B = 1, so neither AND gate, G3 or G4, can produce a 1 output, since one of their inputs will be 0 from the NOT gate, so the sum will be 0. Both inputs to the carry AND gate G5 will be 1, so it will generate a carry output of 1, giving the result of 10.

The half adder is so called because it can sum two binary single-digit

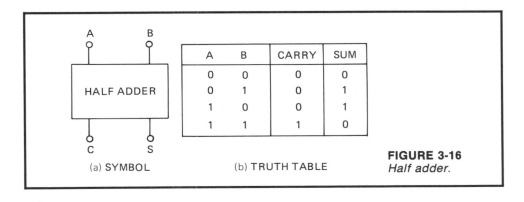

FIGURE 3-16
Half adder.

numbers and pass on the result with a remainder, but it cannot accommodate a third digit carried over from a previous sum. Thus, it is typically used only in the first position in a logical adding chain, where the ones column is certain to be free of another column's remainder. By contrast, a full adder circuit can handle two binary digits plus a carry and may be used anywhere in the chain.

One way to design a full adder is to use two half adders and an OR gate as shown in Figure 3-17. The first half adder sums A and B, and the second adds the resulting sum to a carry input from the previous result to give the final sum. The carry outputs from the two half adders are input to an OR gate, and the output from this gate gives the final carry.

The truth table for the full adder is given in Figure 3-17b. It can be determined by analyzing the full adder circuit that a carry bit cannot exist at the outputs of both half adders simultaneously.

It is interesting to count the number of gates required for these adders. For a single full adder as shown in Figure 3-17, we need 6 AND gates, 3 OR gates and 4 NOT gates, for a total of 13 gates, just to add two binary digits in a single position. Since most microcomputers are designed to handle more than 10-digit decimal numbers, i.e., decimal numbers greater than 10,000,000,000 (2^{33}) and since all computers operate on binary numbers, we need an adder at least 33 digits wide. So, to add two 33-bit binary numbers, we need 32 full adders and a half adder for a total of 422 logic gates. With the requirement

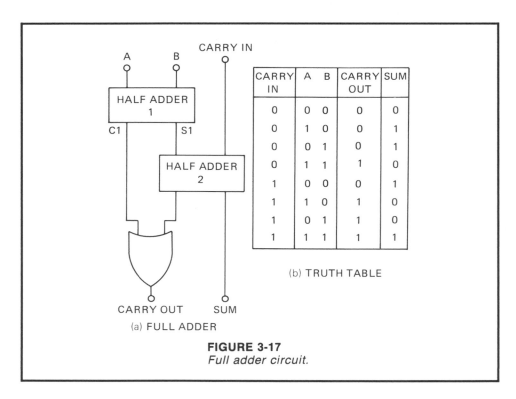

CARRY IN	A	B	CARRY OUT	SUM
0	0	0	0	0
0	1	0	0	1
0	0	1	0	1
0	1	1	1	0
1	0	0	0	1
1	1	0	1	0
1	0	1	1	0
1	1	1	1	1

(b) TRUTH TABLE

(a) FULL ADDER

FIGURE 3-17
Full adder circuit.

of repeated addition to obtain a multiplication, along with the facilities for other arithmetic manipulation, it is easy to understand why the number of gates in the arithmetic logic unit (ALU) of a microprocessor can be up to 10,000.

Timing Devices   Timing devices are used in process industrial control systems to provide a repeatable and predictable sequence of events. Many processes require timing of the events in the system; for example, painting metal parts on an assembly line might require precise timing of the application of each coat of paint.

This section will discuss methods for developing a single pulse and a train of pulses for initiation and timing control in digital systems. The solid-state circuits that perform these tasks are called monostable and astable multivibrators. The monostable device provides a single pulse with adjustable pulse width, while the astable device produces a pulse train with variable duty cycle and frequency.

Astable Multivibrator   The model 555 integrated circuit (IC) is the most popular IC used to produce timing pulses for a digital system. This device requires an external resistor and capacitor network to set the output frequency and the duty cycle. Figure 3-18 gives the IC pin identification as well as the astable multivibrator circuit for a 555.

The clock frequency (f_c) for a 555 astable multivibrator is calculated using the following equation:

$$f_c = 1.44/(R_2 + 2R_1)C \tag{3-18}$$

The duty cycle, which is the ratio of the time spent in the logic zero state to the sum of the time spent in the logic 0 and 1 states, is given by

$$D = R_1/(R_1 + R_2) \tag{3-19}$$

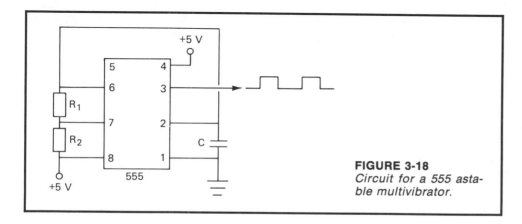

FIGURE 3-18
Circuit for a 555 astable multivibrator.

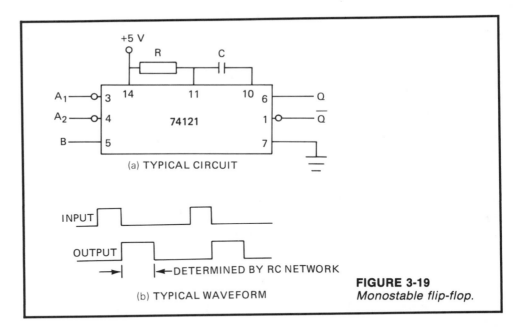

FIGURE 3-19
Monostable flip-flop.

Thus, a duty cycle of 0.5 produces a square wave and is obtained when $R_1 = R_2$.

Monostable Multivibrator   The most common integrated circuit used to produce a single pulse for timing is the 74121. This monostable device, or one-shot, provides precisely timed intervals for use in digital circuits. It is generally used to modify the duration of a digital signal by either stretching (increasing pulse width) or shortening the signal (decreasing pulse width). Figure 3-19 gives a typical circuit for a one-shot using the 74121 along with the circuit waveform.

The pulse width of the output signal is controlled in the circuit by the external feedback components (resistor-capacitor network). If no external feedback components are used, the circuit yields a pulse width output of 30 to 50 ns. The pulse width can be determined from Equation 3-20.

$$T \cong 0.7 \ RC \tag{3-20}$$

where: T = pulse width in seconds
         R = resistance in ohms
         C = capacitance in farads

The manufacturer's data sheet for the 74121 one-shot should be consulted for limitations on the values of R and C.

Latching Devices and Flip-Flops   In the logic gates (AND, OR, NOT, etc.) discussed so far, the output responded immediately to changes at

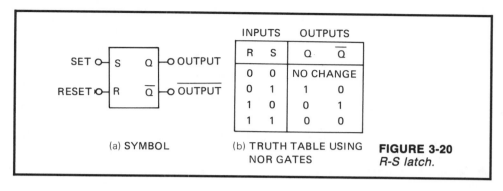

INPUTS		OUTPUTS	
R	S	Q	$\overline{Q}$
0	0	NO CHANGE	
0	1	1	0
1	0	0	1
1	1	0	0

(a) SYMBOL   (b) TRUTH TABLE USING NOR GATES

FIGURE 3-20
R-S latch.

the input. Several digital circuit applications require devices that do not respond to every input logic change.

Latches belong to one class of digital devices whose outputs may or may not follow the input. The latch output does not return to a predetermined logic level but holds its output until the controlling input is altered in an allowable manner.

The flip-flop is another digital device that does not respond to every input change. The difference between a latch and a flip-flop is that a latch is asynchronous to a correct input logic change immediately, while the flip-flop device does not respond to the input logic change until a clock input signal is sent to the device. This section will discuss general characteristics of R-S latches, D latches, and the J-K flip-flop.

R-S Latch   The R-S (Reset-Set) latch is a logic circuit having two possible states. It is designed so that an enabling level on the S (Set) input, with R in the opposite state, yields a Q = high output. An enabling level on the R (Reset) input, with S in the opposite state, yields a Q = low output. Figure 3-20(a) shows the logic symbol for an R-S latch; Figure 3-20(b) gives the truth table for an R-S latch using NOR gates.

EXAMPLE 3-6

Problem: Design an R-S latch using NOR gates only.

Solution: If two-input NOR gates are used, the circuit can be designed as follows:

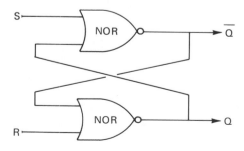

D-Latch   The D-latch is the most commonly used gated latch. It is a bistable device with a single data input (D). This single input is obtained by adding an inverter to the basic R-S circuit inputs to ensure that R and S are always in the opposite state, thus eliminating the possibility of a race condition. A race condition occurs when two inputs are switched at the same time and the resulting output is unpredictable. Race situations need to be avoided if the digital system is to function reliably.

In the D-latch, a change in the data input will be transferred to the output whenever the proper logic level is present on the enable input (G); i.e., as long as the latch is enabled, the output (Q) will follow the input. Figure 3-21 shows the standard symbol and truth table for a D-latch.

J-K Latch   The symbol and truth table for a J-K latch are given in Figure 3-22. The J-K latch is similar to the R-S latch with one exception: If two high inputs occur simultaneously, the J-K latch outputs will reverse their output states. This eliminates the undefined state found in the R-S latch.

J-K Flip-Flop   The symbol and function table for a J-K flip-flop are shown in Figure 3-23. The two basic types of triggering methods employed when using a J-K flip-flop circuit are edge and master-slave. Edge triggering transfers input data to the output on a predetermined clock transition. In master-slave triggering, input data is sampled when the clock input is high and is transferred to the output on the trailing edge of the clock. When using master-slave triggering, the input data must not change during the period of time the clock is high.

The J-K flip-flop is the basic building block of most digital computers. Normally, computers are based on 8-bit, 16-bit, or 32-bit words, and the single bit in the computer word is generally a flip-flop circuit that can assume a value of 0 or 1.

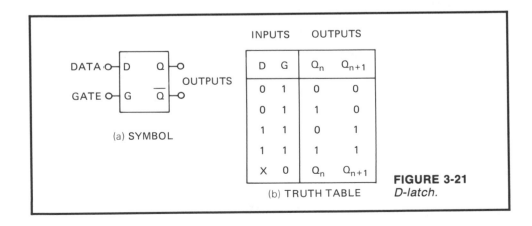

INPUTS		OUTPUTS	
D	G	Q_n	Q_{n+1}
0	1	0	0
0	1	1	0
1	1	0	1
1	1	1	1
X	0	Q_n	Q_{n+1}

(a) SYMBOL

(b) TRUTH TABLE

FIGURE 3-21
D-latch.

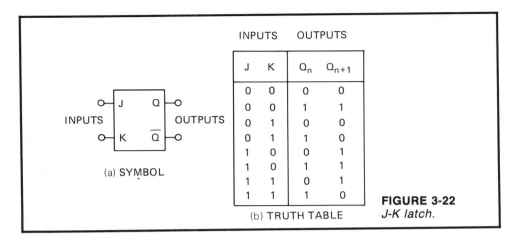

FIGURE 3-22
J-K latch.

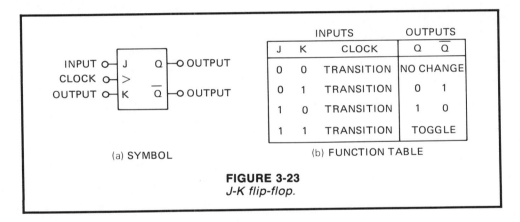

FIGURE 3-23
J-K flip-flop.

EXERCISES

3.1 Verify the logic identity $A + B = B + A$.

3.2 Verify the logic identity $A + AB + 1 = 1$.

3.3 Simplify the logic equation $Z = \overline{C}DA + A\overline{B}D + BC + \overline{A}C$.

3.4 Synthesize the logic function $AC + A\overline{B}D + DC + A\overline{D}$ with two-input NOR and two-input NAND gates.

3.5 Synthesize $A\overline{C} + B \times (C + D)$ with two-input NAND gates only.

3.6 Design a 7-input, odd parity generator using XOR and NOT gates.

3.7 Design a full binary adder circuit to add 4-digit binary numbers.

3.8 Design an astable multivibrator using a 555 IC and a 1 μfarad capacitor to produce a square wave with a frequency of 400 Hz (cycles per second).

3.9 Design an R-S latch using NAND gates.

BIBLIOGRAPHY

1. Kintner, P. M., *Electronic Digital Techniques*, McGraw-Hill Book Company, 1968.

2. Floyd, T. L., *Digital Logic Fundamentals*, Charles E. Merrill Publishing Company, 1977.

3. Malvino, A. P., *Digital Computer Electronics—An Introduction to Microcomputers*, Second Edition, McGraw-Hill Book Company, 1983.

4. Grob, B., *Basic Electronics*, Fifth Edition, McGraw-Hill Book Company, 1984.

5. Budak, A., *Passive and Active Network Analysis and Synthesis*, First Edition, Houghton Mifflin Co., 1974.

6. Boylestad, R. L., and Nashelsky, L., *Electronic Devices and Circuit Theory*, Third Edition, Prentice-Hall, Inc., 1982.

7. Clare, C. R., *Designing Logic Systems Using State Machines*, McGraw-Hill, Inc., 1973.

8. Wickes, W. E., *Logic Design with Integrated Circuits*, John Wiley & Sons, Inc., 1968.

9. Hibberd, R. G., *Integrated Circuits A Basic Course for Engineers and Technicians*, McGraw-Hill Book Company, 1969.

4

Electrical Design

Introduction   The design of programmable controller systems requires knowledge of the basic principles of electrical design. This chapter discusses the fundamentals of electricity, electrical conductors, operation of electrical devices, electrical ladder diagrams and symbols, and typical electrical control applications.

Fundamentals of Electricity   Electricity is a fundamental force in nature that can produce heat, motion, light, and many other physical effects. This force is an attraction or repulsion between electric charges called electrons and protons. An electron is a small particle having a negative electric charge. A proton is a basic particle with a positive charge. It is the arrangement of electrons and protons that determines the electrical characteristics of all substances. As an example, this page of paper has electrons and protons in it, but there is no evidence of electricity because the number of electrons equals the number of protons. In this case, the opposite electrical charges cancel, making the paper electrically neutral.

If we want to use the electrical forces associated with positive and negative charges, work must be done to separate the electrons and protons. For example, an electric battery can do work because its chemical energy separates electric charges to produce an excess of electrons at its negative terminal and an excess of protons at its positive terminal.

The basic terms encountered in electricity are electric charge (Q), current (I), voltage (V), and resistance (R). The charge of a very large number of electrons and protons is required for common applications of electricity. Therefore, it was convenient to define a practical unit called the coulomb (C) as equal to the charge of 6.25×10^{18} electrons or protons stored in a dielectric. The symbol for electric charge is Q, standing for quantity of charge, and a charge of 6.25×10^{18} electrons or protons is stated as Q = 1 C. Voltage is the potential difference in opposite charges. Fundamentally, the volt is a measure of the work needed to move an electric charge. When one joule of work is required to move 6.25×10^{18} electrons between two points, the potential difference is 1 volt (V).

When the potential difference between two different charges forces a third charge to move, the charge in motion is called electric current. If the charge moves at the rate of 6.25×10^{18} electrons per second past a given point, the amount of current is defined as one ampere (A). In equation form $I = Q/t$, where I is the current in amperes, Q is the charge in coulombs, and t is time in seconds.

EXAMPLE 4-1

Problem: A charge of 15 coulombs moves past a given point every second. What is the current flow?

Solution: Since current flow is defined as the amount of charge (Q) that passes a given point per unit of time (t), we have

$$I = \frac{Q}{t} = \frac{15 \text{ C}}{1 \text{ s}}$$

$$I = 15 \text{ A}$$

The fact that a conductor carrying electric current can become hot is evidence that the work done by the applied voltage in producing current must be meeting some form of opposition. This opposition, which limits current flow is called resistance.

Conductivity, Resistivity and Ohm's Law An important physical property of some material is called conductivity, i.e., the ability to pass electric current. Suppose we have an electric wire (conductor) of length L and cross-sectional area A, and we apply a voltage V between the ends of the wire. If V is in volts and L is in meters, we can define the voltage gradient, E, as

$$E = \frac{V}{L} \text{ (volts/meter)} \tag{4-1}$$

If a current I in amperes flows through a wire of area A in meters^2, we can define the current density J as

$$J = \frac{I}{A} \text{ (amps/meters}^2) \tag{4-2}$$

The conductivity (C) is defined as the current density per unit voltage gradient (E) or, in equation form,

$$C = \frac{J}{E} \text{ (amps/meter}^2)/(\text{volts/meter})$$

(4-3)

$$\text{or } C = \frac{I/A}{V/L} = \frac{I}{V} \times \frac{L}{A}$$

Resistivity (ρ) is defined as the inverse of conductivity, or

$$\rho = \frac{1}{C}$$

The fact that resistivity is a natural property of certain materials leads to the basic principle of current flow called Ohm's Law.

Consider a wire of length (L) and area (A). If it has resistivity, ρ, then its resistance (R) is

$$R = \rho \frac{L}{A}$$

(4-4)

The units of resistance are ohms. Since the resistivity, ρ, is the reciprocal of conductivity, we obtain the following:

$$\rho = \frac{V}{I} \times \frac{A}{L}$$

(4-5)

When Equation 4-5 is substituted into Equation 4-4, we obtain

$$R = \frac{V}{I}$$

(4-6)

This relationship $R = V/I$ or $V = IR$ is called Ohm's Law, which assumes that the resistance of the material used to carry the current flow is linear, i.e., if the voltage across the resistance is doubled, the current through it also doubles. The resistance of materials like carbon, aluminum, copper, silver, gold, and iron is linear.

Wire Resistance   To compare the resistance and size of one conductor with another, a standard or unit size of conductor was established. A convenient unit of linear measurement, as far as the diameter of a section of wire is concerned, is the mil (0.001 of an inch); and a convenient unit of wire length is the foot. The standard unit of size in most cases is the *mil-foot*; that is, a wire will have unit size if it has a diameter of 1 mil and a length of 1 foot.

The *circular mil* is the standard unit of wire cross-sectional area used in US wire tables. Because the diameter of round wires used to conduct electricity may be only a small fraction of an inch, it is convenient to express these

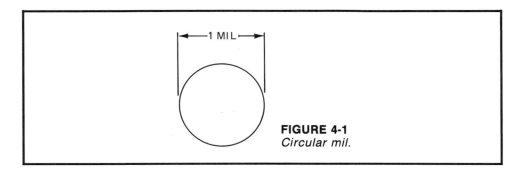

FIGURE 4-1
Circular mil.

diameters in mils, to avoid the use of decimals. For example, the diameter of a conductor is expressed as 10 mils instead of 0.010 inch. A circular mil is the area of a circle having a diameter of 1 mil, as shown in Figure 4-1.

A *circular-mil per foot* is a unit conductor 1 foot in length having a cross-sectional area of 1 circular mil. Because it is considered a unit conductor, the cmil per foot is useful in making comparisons of the resistivity of various materials.

EXAMPLE 4-2

Problem: Calculate the area in circular mils of a wire with a diameter of 0.004 in.

Solution: First, we must convert the diameter to mils. Since we have defined 0.001 in. = 1 mil, this implies that 0.004 in. = 4 mils. So the circular mil area is $(4 \text{ mil})^2$ or 16 cmil.

The specific resistances in ohms-cmils per foot of some common substances at 20°C are given in Table 4-1. We can use the specific resistance (ρ) given in Table 4-1 to find the resistance of a conductor of length L and cross-sectional area A by using Equation 4-4, $R = \rho L/A$.

TABLE 4-1
Specific Resistance (ρ) at 20°C

Substance	cmil-ohms per foot
Silver	9.8
Copper (drawn)	10.37
Gold	14.7
Aluminum	17.02
Tungsten	33.2
Brass	42.1
Steel (soft)	95.8
Nichrome	660.0

EXAMPLE 4-3

Problem: Find the resistance of 100 feet of copper wire having a cross-sectional area of 10,400 circular mils; the wire temperature is 20°C.

Solution: The specific resistance of copper wire from Table 4-1 is 10.37 cmil-ohms/ft. Substituting the known values in Equation 4-4, the resistance, R, is determined as follows:

$$R = \rho \frac{L}{A}$$

$$= 10.37 \text{ cmil-ohms/ft} \times \frac{100 \text{ ft}}{10400 \text{ cmil}}$$

$$= 0.1 \ \Omega$$

The equation for the resistance of a conductor, $R = \rho \dfrac{L}{A}$, can be used in many applications. For example, if R, ρ, and A are known, the length can be determined by a simple mathematical transposition. A typical example is locating a problem ground point in a telephone line. Specially designed test equipment is used, which operates on the principle that the resistance of a line varies directly with length. Thus, the distance between the test point and a fault can be computed accurately.

EXAMPLE 4-4

Problem: The resistance to ground on a faulty underground telephone line is 5 ohms. Calculate the distance to the point where the wire is shorted to ground, if the line is a copper conductor with a diameter of 1020 cmils and the ambient temperature is 20°C.

Solution: To calculate the distance to the point where the wire is shorted to ground, we use L = R × A/ρ. Since R = 5 Ω, A = 1020 cmils, and ρ = 10.37 cmil-Ω/ft, we have

$$L = \frac{R \times A}{\rho} = \frac{5 \ \Omega \times 1020 \text{ cmil}}{10.37 \text{ cmil-}\Omega\text{/ft}}$$

$$L = 492 \text{ feet}$$

Standard Wire Gage Sizes Electrical conductors are manufactured in sizes numbered according to a system known as the American Wire Gage (AWG). As can be seen in Table 4-2, the wire diameters become smaller as the gage numbers increase. The largest wire size listed in the table is 4 and the smallest is 24. This is the normal range of wire sizes encountered in programmable controller applications. The complete AWG table goes from 0000 to 40; the

TABLE 4-2
Standard Annealed Solid Copper Wire
(American Wire Gage)

AWG no.	Diameter (mils)	Cross section (circular mils)	Ohms per 1000 feet at 25°C	at 65°C
4	204.0	41700.0	0.253	0.292
6	162.0	26300.0	0.403	0.465
8	128.0	16500.0	0.641	0.739
10	102.0	10400.0	1.02	1.18
12	81.0	6530.0	1.62	1.87
14	64.0	4110.0	2.58	2.97
16	51.0	2580.0	4.09	4.73
18	40.0	1620.0	6.51	7.51
20	32.0	1020.0	10.4	11.9
22	25.3	642.0	16.5	19.0
24	20.1	404.0	26.2	30.2

larger and smaller sizes not listed in Table 4-2 are manufactured but not commonly used in programmable controller application.

EXAMPLE 4-5

Problem: Determine the resistance of 2500 feet of 14 AWG copper wire. Assume an ambient temperature of 25°C.

Solution: Using Table 4-2, we see that 14 AWG wire has a resistance of 2.58 Ω per 1000 ft at 25°C. So the resistance of 2500 ft is calculated as follows:

$$R = \frac{2.58 \ \Omega}{1000 \ \text{ft}} \times 2500 \ \text{ft} = 6.5 \ \Omega$$

Direct Current and Alternating Current   Basically, two types of voltages are encountered in PC systems; direct current (dc) and alternating current (ac). In direct current, the flow of charges is in just one direction. A battery is an example of a dc power source (see Figure 4-2). A graph of dc voltage versus time is shown in Figure 4-2a, and the symbol for a battery is shown in Figure 4-2b.

An alternating voltage source periodically reverses or alternates in polarity. Therefore, the resulting alternating current periodically reverses in direction (see Figure 4-3a). The symbol for an ac power source is shown in Figure 4-3b.

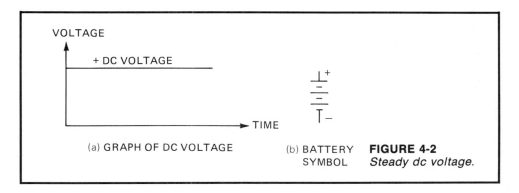

(a) GRAPH OF DC VOLTAGE

(b) BATTERY SYMBOL

FIGURE 4-2
Steady dc voltage.

The 60-cycle ac power used in homes and industry is a common example of ac voltage. This frequency means that the voltage polarity and current direction go through 60 cycles of reversal per second. The unit for 1 cycle per second is 1 hertz (Hz). Therefore, 60 cycles per second is a frequency of 60 Hz.

Series Resistance Circuits When the components in a circuit are connected in successive order with the end of a component joined to the end of the next element, they form a series circuit. An example of a series circuit is shown in Figure 4-4.

In this circuit, the current (I) flows from the power supply through the two resistors (R_1 and R_2) and back to the power supply. According to Ohm's Law, the amount of current (I) flowing between two points in a circuit equals the potential difference (V) divided by the resistance (R) between these points. If V_1 is the voltage drop across R_1, V_2 is defined as the voltage drop across R_2 and the current (I) flows through both R_1 and R_2, we obtain from Ohm's Law:

$$V_1 = IR_1 \text{ and } V_2 = IR_2$$

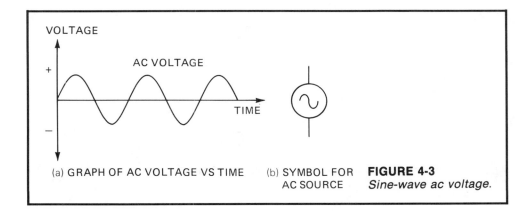

(a) GRAPH OF AC VOLTAGE VS TIME

(b) SYMBOL FOR AC SOURCE

FIGURE 4-3
Sine-wave ac voltage.

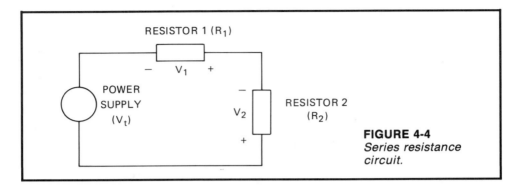

FIGURE 4-4
Series resistance circuit.

Therefore,

$$V_t = IR_1 + IR_2$$

If we divide both sides of the equation by I, we obtain

$$V_t/I = R_1 + R_2$$

Since $V_t/I = R_t$, we obtain the total resistance of the series circuit of Figure 4-4 as

$$R_t = R_1 + R_2$$

We can derive a more general equation for any number of resistors in series by using the classical law of conservation of energy. According to this law, the energy or power supplied to a series circuit must equal the power dissipated in the resistors in the circuit. Thus,

$$P_t = P_1 + P_2 + P_3 + \ldots + P_n$$

or, since $P = I^2R$,

$$I^2R_t = I^2R_1 + I^2R_2 + I^2R_3 + \ldots + I^2R_n$$

If we divide this equation by I^2, we obtain the series resistance formula,

$$R_t = R_1 + R_2 + R_3 + \ldots + R_n \qquad (4\text{-}7)$$

EXAMPLE 4-6

Problem: Assume that the power supply voltage in the series circuit of Figure 4-4 is 24 V dc and that resistor $R_1 = 250 \ \Omega$ and resistor $R_2 =$

250 Ω. Find the total current flow (I_t) in the circuit and the voltage drop across R_1 and R_2.

Solution: To calculate the circuit current I_t, first find the total circuit resistance R_t using Equation 4-7:

$$R_t = R_1 + R_2$$
$$= 250\ \Omega + 250\ \Omega$$
$$= 500\ \Omega$$

Now, according to Ohm's Law:

$$I_t = \frac{V_t}{R_t}$$
$$= \frac{24\ V}{500\ \Omega}$$
$$= 48\ mA$$

The voltage across R_1 (i.e., V_1) is given by

$$V_1 = R_1 I_t$$
$$= (250\ \Omega)(48\ mA)$$
$$= 12\ V$$

The voltage across R_2 (i.e., V_2) is obtained as follows:

$$V_2 = R_2 I_t$$
$$= (250\ \Omega)(48\ mA)$$
$$= 12\ V$$

Parallel Resistance Circuits   When two or more components are connected across a power source, they form a parallel circuit. Each parallel path is called a branch, and each has its own current. In other words, parallel circuits have one common voltage across all the branches, but individual branch currents. These characteristics are opposite from series circuits, which have one common current but individual voltage drops.

A parallel circuit with two resistors across a power source is shown in Figure 4-5. The total resistance (R_t) across the power supply can be found by Ohm's Law as follows: Divide the voltage across the parallel resistances by the total current of all the branches. In the circuit of Figure 4-2, $I_t = I_1 + I_2$

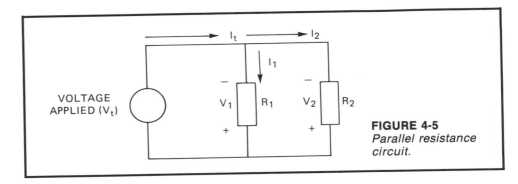

FIGURE 4-5
Parallel resistance circuit.

and $I_1 = V_t/R_1$, $I_2 = V_t/R_2$, so that $I_t = V_t/R_1 + V_t/R_2$. Since $I_t = V_t/R_t$, we obtain

$$\frac{1}{R_t} = \frac{1}{R_1} + \frac{1}{R_2}$$

(4-8)

or

$$R_t = \frac{R_1 \times R_2}{R_1 + R_2}$$

We can derive the general reciprocal resistance formula for any number of resistors in parallel from the fact that the total current I_t is the sum of all the branch currents, or

$$I_t = I_1 + I_2 + I_3 + \ldots + I_n$$

This is simply the law of electrical charge conservation,

$$dQ_t/dt = dQ_1/dt + dQ_2/dt + dQ_3/dt + \ldots dQ_n/dt$$

This law implies that, since no charge accumulates at any point in the circuit, the charge dQ_t from the power source in time dt must appear as charges dQ_1, dQ_2, dQ_3, ..., dQ_n through the resistors at the same time.

To illustrate the basic concepts of a parallel circuit, let's look at an example problem.

EXAMPLE 4-7

Problem: Assume that the voltage for the parallel circuit shown in Figure 4-5 is 12 V and we need to find the currents I_t, I_1 and I_2, and the parallel resistance R_t, given that $R_1 = 10$ kΩ and $R_2 = 10$ kΩ.

Solution: We can find the parallel circuit resistance R_t by using Equation 4-8:

$$R_t = \frac{R_1 \times R_2}{R_1 + R_2}$$

$$R_t = \frac{(10 \text{ k})(10 \text{ k})}{10 \text{ k} + 10 \text{ k}} \Omega = 5 \text{ k}\Omega$$

The total current flow is given by

$$I_t = \frac{V_t}{R_t} = \frac{12 \text{ V}}{5 \text{ k}\Omega} = 2.4 \text{ mA}$$

The current in branch 1 (I_1) is obtained as follows:

$$I_t = \frac{V_t}{R_1} = \frac{12 \text{ V}}{10 \text{ k}\Omega} = 1.2 \text{ mA}$$

Since $I_t = I_1 + I_2$, the current flow in branch 2 is given by

$$I_2 = I_t - I_1 = 2.4 \text{ mA} - 1.2 \text{ mA} = 1.2 \text{ mA}$$

Selection of Wire Size   Several factors must be considered in selecting the wire size in an electrical design application. One factor is the permissible power loss ($P = I^2R$) in the line. This power loss is electrical energy being converted to heat. If the heat produced is excessive, damage to the conductors or system components might result. The use of large conductors will reduce the circuit resistance and, therefore, the I^2R loss. However, large conductors are more expensive than smaller ones, and they are more costly to install.

A second factor is the voltage drop in the circuit. If the resistance of a conductor is excessive, large voltage drops will occur in the transmission of electrical signals. For example, let's assume we are sending a 20 mA dc signal from a remote location to a control panel as shown in Figure 4-6.

If we are using 20 AWG wire, we can easily calculate the wire resistance

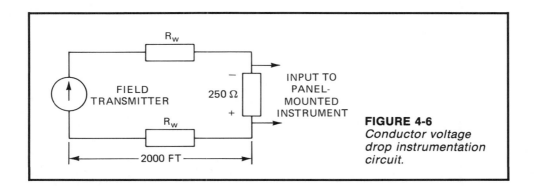

FIGURE 4-6
Conductor voltage drop instrumentation circuit.

(R_w) as follows:

$$R_w = r \frac{L}{A}$$

$$= (10.37 \text{ cmil-}\Omega\text{/ft}) \frac{2000 \text{ ft}}{1020 \text{ cmil}}$$

$$= 20.3 \ \Omega$$

The total resistance of the signal cable is 40.6 Ω, so the load on the field transmitter will be 250 Ω plus 40.76 Ω or 290.6 Ω. This places an additional load on the power supply used by the field transmitter.

A third factor is the current-carrying ability of the line. When current flows in a wire, heat is generated. The temperature of the wire will rise until the heat radiated, or otherwise dissipated, is equal to the heat generated by the current in the wire or cable. If the wire is insulated, the heat produced is not so readily removed as it would be if the conductor were not insulated. So, to protect the insulation from excessive heat, the current through the wire must be kept below a certain value. Rubber insulation will start to deteriorate at relatively low temperatures. Teflon™ and certain plastic insulations retain their insulating properties at higher temperatures, and insulations such as asbestos are effective at still higher temperatures.

Electrical cables might be installed in locations where the ambient temperature is relatively high. In these cases, the heat produced by external sources adds to the total electrical cable heating. Therefore, allowances must

TABLE 4-3
Allowable Carrying Capacities of Conductors
(National Electrical Code) (1)

Size AWG	Rubber insulation, amperes	Varnished cambric insulation, amperes	Other insulations and bare conductors, amperes
18	3	—	6
16	6	—	10
14	15	18	20
12	20	25	30
10	25	30	35
8	35	40	50
6	50	60	70
4	70	85	90
2	90	110	125
0	125	150	200

be made in designing electrical systems for the ambient heat sources. The maximum allowable operating temperature of insulated wires and cables is specified in electrical design tables. For example, the allowable current-carrying capacities of copper wires at not over 30°C (86°F) temperature for single conductors in air are given in Table 4-3 (Reference 1). The ratings listed in Table 4-3 are those permitted by the National Electrical Code for flexible cords and for interior wiring of houses, hotels, office buildings, industrial plants, and other buildings.

This table can be used to determine the safe and proper wire size in an electrical application. An example problem will help to illustrate a typical wire sizing calculation.

EXAMPLE 4-8

Problem: Calculate the correct wire size for the programmable controller system shown.

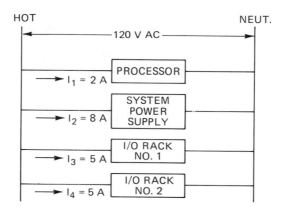

Solution: The total current I_t is the sum of all the branch currents, or

$$I_t = I_1 + I_2 + I_3 + I_4$$

$$I = 2\,A + 8\,A + 5\,A + 5\,A$$

$$= 20\,A$$

Using Table 4-3, we can see that the main power conductors (hot, neutral, and ground) would have to be a minimum of 12 AWG if we use a rubber insulated cable.

Electrical Devices and Symbols   A wide variety of electrically operated devices and instruments are found in programmable controller applications. We will cover the most common devices in this section and give the symbols for each device or instrument.

Electrical Relays   The most common device encountered in electrical control is the electromechanical relay. It is called electromechanical because it consists of an electrically operated solenoid and mechanical contacts. The solenoid is a long wire wound in a close-packed helix; when an electric current is applied, a strong magnetic field is produced. The magnetic force produced moves an iron armature that is used to open or close the mechanical contacts when electrical power is applied. The contacts in the relay make or break connections in electric circuits. A typical relay is shown in Figure 4-7, with two sets of contacts, normally closed (NC) and normally open (NO) with no voltage applied to the relay solenoid.

The electrical schematic symbol for a normally open (NO) contact is given in Figure 4-8a and the symbol for a normally closed (NC) contact is given in Figure 4-8b. The standard symbol for a relay is a circle, as shown in Figure 4-8c. The symbol also includes letter(s) and number(s) to identify the control relay, such as R1, CR10, and so on.

Electric Solenoid Valves   Another common electrically operated device encountered in process control is the solenoid valve. The solenoid valve is a combination of two basic functional units: a solenoid (electromagnet) with its core and a valve body containing one or more orifices. Flow through an orifice is allowed or prevented by the action of the core when the solenoid is energized or deenergized.

Solenoid valves normally have a solenoid mounted directly on the valve body. The core is enclosed and free to move in a sealed tube called a core tube, thus providing a compact assembly. The electrical symbols for a solenoid valve are given in Figure 4-9.

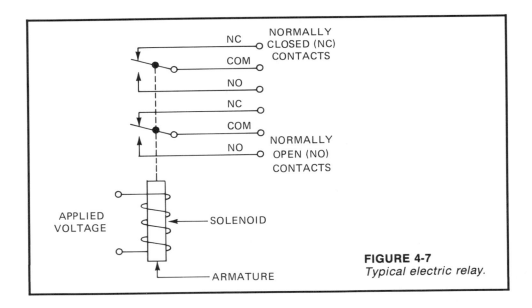

FIGURE 4-7
Typical electric relay.

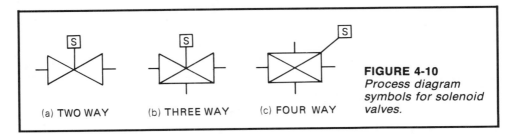

FIGURE 4-8
Electrical relay symbols.

(a) NORMALLY OPEN (NO) CONTACTS

(b) NORMALLY CLOSED (NC) CONTACTS

(c) COIL

FIGURE 4-9
Schematic symbol for solenoid valve.

OR

The three common types of solenoid valves are: (1) direct-acting, (2) internal pilot-operated, and (3) manual reset. In direct-acting valves, the solenoid core directly opens or closes the orifice, depending upon whether the solenoid is energized or deenergized. The valve will operate from zero psi to its maximum rated pressure. The force required to open the valve is proportional to the orifice size and the pressure drop. As the orifice size increases, the force needed increases. To keep the solenoid size small and to open large orifices, internal pilot-operated valves are used.

Internal pilot-operated valves have a pilot and bleed orifice and use the line pressure for operation. When the solenoid is energized, the core opens the pilot orifice and relieves pressure from the top of the valve piston or diaphragm to the outlet side of the valve. The imbalance of pressure causes the line pressure to lift the piston or diaphragm off the main valve orifice, opening the valve. When the solenoid is deenergized, the pilot orifice is closed, and the full line pressure is applied to the top of the piston or diaphragm through the bleed orifice to close the valve.

The manual reset valve must be manually positioned (latched). It will return to its original position when the solenoid is energized or deenergized, depending on the valve construction.

Three basic solenoid valve configurations are available for process control: (a) two-way, (b) three-way and (c) four-way (see symbols used in process diagrams in Figure 4-10). Two-way valves have one outlet and one inlet pipe

(a) TWO WAY

(b) THREE WAY

(c) FOUR WAY

FIGURE 4-10
Process diagram symbols for solenoid valves.

connection. They are available in either normally closed or normally open construction. Three-way solenoid valves have three pipe connections and two orifices (one orifice is always open and one is always closed). These valves are normally used in process control to alternately apply air pressure to and exhaust pressure from a diaphragm-operated control valve or a single-acting cylinder. Four-way solenoid valves are normally used in process control to operate double-acting cylinders. These valves have four pipe connections: one pressure, two cylinders, and one exhaust. In one valve position, pressure is connected to one cylinder port; the other is connected to exhaust. In the other valve state, pressure and exhaust are reversed at the cylinder connections.

Contactors Contactors are devices used to repeatedly make and break electrical power circuits. The contactor can be energized by a low amperage programmable controller output module. The contactor is then used to control higher power loads such as heaters, transformers, and capacitors.

The two types of contactors used with programmable controllers are electronic and magnetic. Magnetic contactors are activated by electromechanical means.

An electric solenoid is the principal operating mechanism for a magnetic contactor. Instead of opening or closing a valve, the linear action of the solenoid is used to open or close sets of contacts.

Other Common Control Devices Hand switches, push buttons, and lights are some of the common control devices encountered in electrical ladder design.

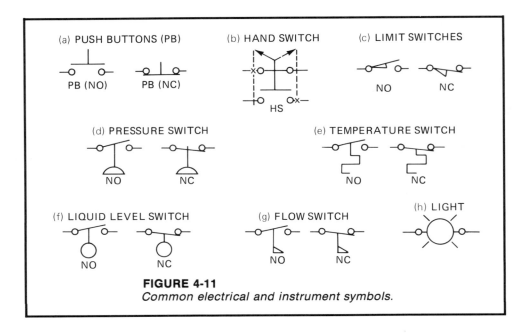

FIGURE 4-11
Common electrical and instrument symbols.

These devices are used to control other devices, such as solenoid valves, electric motors and motor starters, heaters, and so on. Another class of device found on electrical and instrument drawings are field-mounted switches for pressure, temperature, liquid level, and flow sensing.

To make the reading of ladder diagrams easier, a set of standardized symbols has been developed by the American National Standards Institute (ANSI). The most common symbols used in ladder diagrams are shown in Figure 4-11; they are based on ANSI Y32.2 for electric controls.

Ladder Diagrams
Ladder diagrams are a traditional method for describing electrical control systems. These circuits are called ladder diagrams because they look like ladders with rungs. Each rung of a ladder is numbered in order that we can cross reference between sections in the drawings that describe the control.

We can appreciate the utility of ladder diagrams by reviewing a simple control example. A typical electrical ladder control application might be to control the liquid level in a tank (see Figure 4-12). In this application, let's assume the flow into a tank is random, and we need to control the level in the tank by opening or closing an outlet solenoid valve (EV-1). We will also provide the operator with a hand switch to manually turn on the solenoid valve or select automatic level control.

The ladder diagram for this application is shown in Figure 4-13. The control consists of an automatic (AUTO) and manual (MAN) switch, so that the tank level can be controlled automatically using the level switch or manually by a process operator.

A more complex application might be to control tank level between two level switches, a level switch high (LSH) and a level switch low (LSL). In this example, a pump is turned on and off to maintain the liquid level between the two switches. The process shown in Figure 4-14 uses steam at a regulated flow rate to boil down liquid to produce a more concentrated solution, which is drained off periodically. A diamond symbol is used on the process diagram to indicate the pump is interlocked with the level switches on the tank.

The electrical ladder diagram used to control the feed pump and hence the liquid level in the process tank is shown in Figure 4-15.

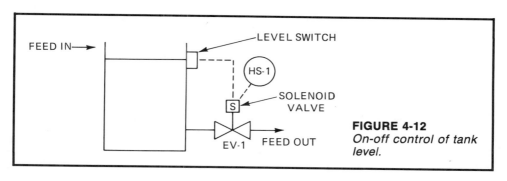

FIGURE 4-12
On-off control of tank level.

FIGURE 4-13
Ladder diagram for tank level control.

To explain the operation of the control circuit shown in the ladder diagram of Figure 4-15, we will assume the tank is empty, the low level switch is closed (made) and the high level switch is closed. When a level switch is energized, the normally open contacts are closed and the normally closed contacts are opened. To start the system, the operator depresses the start push button (PB1). This energizes relay CR1, which seals in the start push button with the first set of contacts (CR1(1)). At the same time, control relay 2 (CR2) is energized in rung 3 through contacts of LSH, CR1(2), and LSL. This turns on the pump motor starter relay K1 using the second set of relay contacts (CR2(2)). The first set of contacts on CR2 is used to seal in the low level switch (LSL), so that when the level in the tank rises above the position of the low level switch the pump will remain ON. The pump will remain ON until the

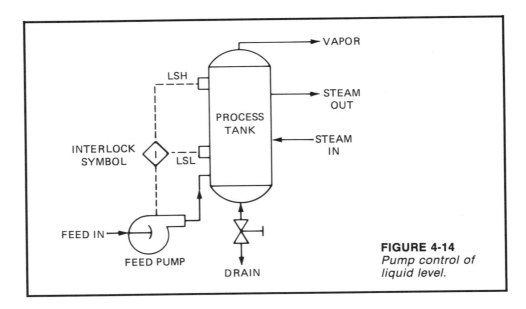

FIGURE 4-14
Pump control of liquid level.

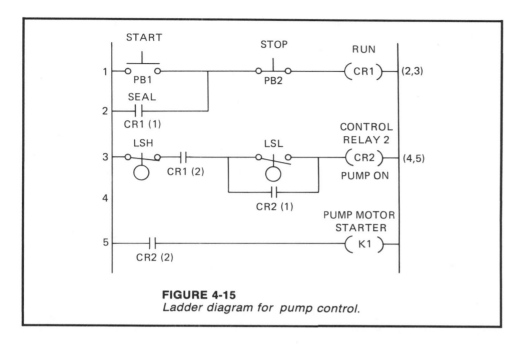

FIGURE 4-15
Ladder diagram for pump control.

level in the tank reaches the high level switch. After the LSH is activated, the pump is turned off. The system will then cycle ON and OFF between the two level switches, until the stop push button is pushed.

EXERCISES

4.1 A charge of 5 coulombs moves past a given point every second. What is the current flow?

4.2 Calculate the current in a conductor, if 25×10^{18} electrons pass a given point in the wire every 2 seconds.

4.3 Calculate the resistance of 500 feet of silver wire with a diameter of 0.001 inch.

4.4 The resistance to ground on a faulty underground telephone line is 15 ohms. Calculate the distance to the point where the wire is shorted to ground, if the line is a 22 AWG copper conductor and the ambient temperature is 25°C.

4.5 Calculate the current flow in a series circuit with a 48-V dc power supply and four resistors with the following values: 250 Ω, 500 Ω, 500 Ω, and 250 Ω.

4.6 Find currents I_1, I_2, I_3, I_{23}, and I_t in the circuit shown when the applied voltage (V_t) is 120 V ac and 20 V dc.

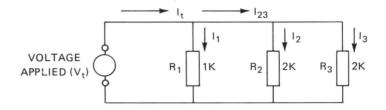

4.7 Calculate the wire size needed for the programmable controller system shown.

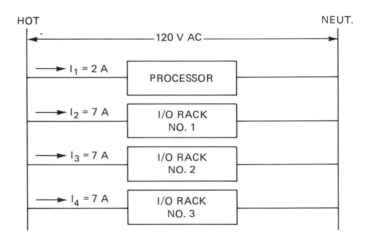

4.8 Explain the operation of the three common types of electrically operated solenoid valves.

4.9 Design an electrical control system using relays to automatically fill and empty a liquid storage tank under the following process conditions:

1. An electrically operated pump is used to fill a tank when a low level is reached.

2. Liquid is removed from the tank using an electrically operated solenoid valve under operator control.

3. The storage tank is automatically transferred to another storage tank if a high level is reached and the pump turns off on tank high level.

REFERENCE

1. Weast, R. C., (Ed.) and Astle, M. J. (Asst. Ed.); *CRC Handbook of Chemistry and Physics*, CRC Press, Inc., 59th Edition, 1978–1979.

BIBLIOGRAPHY

1. Hughes, T. A., *Measurement and Control Basics*, Instrument Society of America Publications, 1988.

2. Malvino, A. P., *Electronics Principles*, Second Edition, McGraw-Hill Book Co., 1979.

3. Budak, A., *Passive and Active Network Analysis and Synthesis*, Houghton Mifflin Co., 1974.

4. Liptak, B. G., (Ed.) and Venczel, K., (Asst. Ed.); *Process Control Instrument Engineers Handbook*, Chilton Book Company, Revised Edition, 1985.

5

Input/Output System

Introduction   The input/output (I/O) system provides the physical connection between the process equipment and the central processing unit (CPU) or simply "processor". It uses various interface circuits and/or modules to sense and measure physical quantities of the process, such as motion, level, temperature, pressure, current, and voltage. Based on the status sensed or values measured, the processor controls various devices, such as valves, motors, pumps, and alarms, to exercise control over a machine or process using output modules.

These modules are mounted in equipment housings as shown in Figure 5-1. In most programmable controller systems any module can be inserted into any I/O slot, and the housings are designed so that the I/O modules can be removed without turning off the ac power or removing the field wiring. Most I/O modules use printed circuit board technology, and the circuit boards have an edge connector that can be inserted into a plug in the backplane connector of the rack. This backplane has a printed circuit card that contains the parallel communications bus to the processor and the dc voltages to operate the logic on the I/O modules.

Discrete Input/Output Modules   Discrete is the most common class of input/output in a PC system. These types of interface modules connect field input devices, which will provide an input signal that is separate and distinct in nature, or field output devices that will require a separate and distinct signal to control their state. This characteristic limits the discrete I/O interfaces to sensing signals that are on/off, open/closed, or equivalent to a switch closure. To the interface module, all discrete inputs are essentially devices with two states. Likewise, output control is limited to devices that require switching only one of two states, such as on/off, open/closed, or ex-

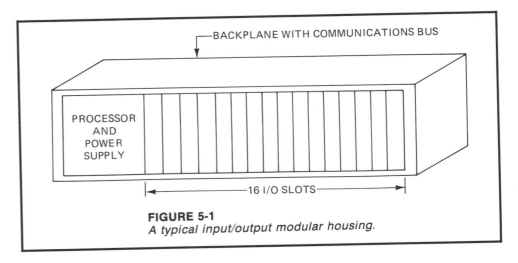

FIGURE 5-1
A typical input/output modular housing.

tended/retracted. Table 5-1 lists the most common discrete input/output devices in process control applications.

Each discrete input and output is powered by some field-supplied voltage source (e.g., 120 V ac, 24 V dc). I/O interface circuits are available at various ac and dc voltage ratings, as listed in Table 5-2.

When in operation, if an input switch is closed, the input module senses the supplied voltage and converts it to a logic-level signal acceptable to the CPU to indicate the status of that device. A logic 1 indicates ON or CLOSED, and a logic 0 indicates OFF or OPENED. In operations the output interface circuit switches the supplied control voltage that will energize or deenergize the device. If an output is turned ON through the control program, the supplied control voltage is switched by the interface circuit to activate the referenced (addressed) output device.

TABLE 5-1
Typical Discrete I/O Field Devices

Input Field Devices	Output Field Devices
Selector switches	Annunciators
Push buttons	Electric control relays
Photoelectric cells	Electric fans
Limit switches	Lights
Logic gates	Logic gates
Proximity switches	Alarm horns
Process switches (level, flow, etc.)	Motor starters
Motor starter contacts	Electric valves
Control relay contacts	Alarm lights

TABLE 5-2
Typical Discrete I/O Signal Types

I/O Signal types	I/O Signal Types
5 volt dc	120 volt ac/dc
12 volt dc	230 volt ac/dc
24 volt ac/dc	Relay contacts
48 volt ac/dc	100 volt dc

AC Input Modules   A block diagram of a typical ac input module is shown in Figure 5-2. Input modules vary widely among PC manufacturers, but in general all ac input modules operate in a manner similar to that described in this diagram. The input circuit is composed of two primary parts: the power section and the logic section. The power and logic sections of the circuit are normally coupled with a circuit that electrically isolates the input power section from the logic circuits. This electrical isolation is very important in a normally noisy industrial environment. The main problem with the early application of computers to process control is that the inputs and outputs were not designed for the harsh industrial environment.

The power section of an input module basically performs the function of converting the incoming voltage (115 V ac, 230 V ac, etc.) from an input device, such as those described in Table 5-1, to a logic-level signal to be used by the processor during its program control scan. The bridge rectifier circuit converts the incoming signal (ac or dc) to a dc level that is sent to a filter circuit, which protects against electrical noise on the input power line. This filter causes a signal delay that is typically 10–25 msec. The threshold circuit detects whether the incoming signal has reached the proper voltage level for the specified input

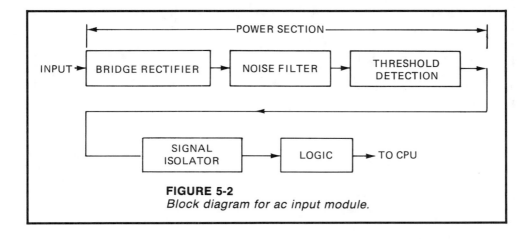

FIGURE 5-2
Block diagram for ac input module.

rating. If the input signal exceeds and remains above the threshold voltage for a duration of at least the filter delay, the signal will be accepted as a valid input.

When a valid signal has been detected, it is passed through the isolation circuit, which completes the electrically isolated transition from ac to logic level. The dc signal from the isolator is used by the logic circuit and made available to the processor via the data communications bus. Electrical isolation is provided so that there is no electrical connection between the field device (power) and the controller (logic). This electrical separation helps prevent large voltage spikes from damaging the logic side of the interface (or the controller). The coupling between the power and logic sections is normally provided by an optical coupler or a pulse transformer. This electrical isolation is one of the main reasons the programmable controller has gained very wide acceptance in the process industries.

Most input modules will have a power indicator to signify that the proper input voltage level is present (a switch is closed). A light-emitting diode (LED)

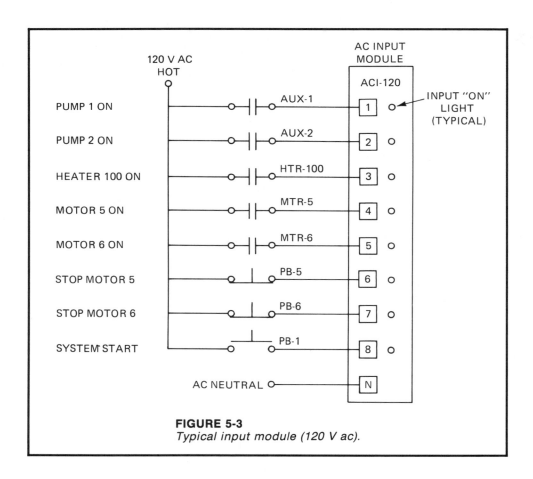

FIGURE 5-3
Typical input module (120 V ac).

indicator may also be available to indicate the status of the input. These indicating lights are an important aid during system start-up and troubleshooting. An ac input connection diagram is shown in Figure 5-3.

The letters "ACI-120" are used to designate a 120-V ac input module. This designation will be used in this book to simplify wiring diagrams and system drawings. We will also use "ACI-220" to designate a 220-V ac input module.

Direct Current (dc) Input Modules   The dc voltage input modules convert discrete on/off direct current inputs to logic level signals compatible with the programmable controller. They are generally available in three voltage ranges: 12, 24, and 48 volts dc. Typical instruments that are compatible with the module include limit switches, valve position switches, push buttons, dc proximity switches, float switches, and photoelectric sensors. Figure 5-4 shows some typical field inputs to a 24-V dc input module.

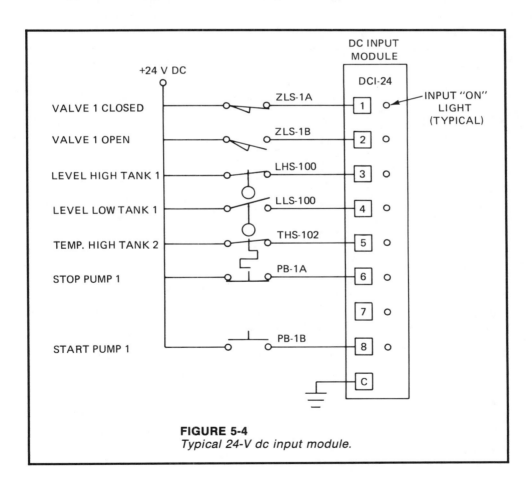

FIGURE 5-4
Typical 24-V dc input module.

The model number for the 24 V dc input module is DCI-24 in this presentation, for convenience. We will also designate a 12-V dc input module as "DCI-12" and a 48-V dc input module as "DCI-48". All three modules will have the wiring diagram shown as the 24-V dc module in Figure 5-4.

Transistor-Transistor Logic (TTL) Input Modules   The TTL input modules allow the controller to accept signals from TTL-compatible devices, including solid-state controls and sensing instruments. TTL inputs are also used for interfacing with some 5-V dc level control devices and several types of photoelectric sensors. The TTL interface has a configuration similar to the dc input modules; however, the input delay time caused by filtering is generally much shorter. TTL input modules normally require an external +5-V dc power supply. Figure 5-5 shows a typical TTL input connection diagram.

Isolated Input Module   Input and output modules usually have a common return line connection for each group of inputs or outputs on a single module. However, sometimes it may be required to connect an input device of different

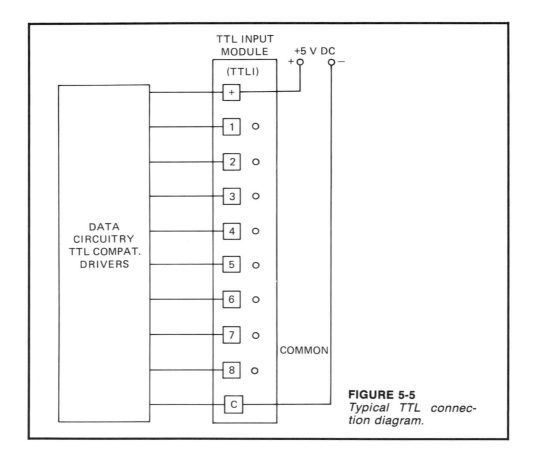

FIGURE 5-5
Typical TTL connection diagram.

ground levels to the controller. In this case, isolated input modules with separate return lines for each input circuit are available (ac or dc) to accept these signals. The operation of the isolated interface is the same as the standard discrete I/O, except the common of each input is separated from the other commons in the module. The result is that the isolated input module requires twice as many input connection terminals, so it can accommodate only half the inputs in the same physical space as shown in Figures 5-6. The module number for the isolated 120-V ac input module is selected as "IACI-120" and the isolated 220-V ac input module can be designated as "IACI-220".

AC Output Modules Figure 5-7 shows a block diagram of a typical ac output module. AC output modules vary widely among PC manufacturers. The block diagram, however, describes basic operation of most ac output modules. The circuit consists primarily of the logic and power sections, coupled by an isolation circuit. The output interface can be thought of as a simple switch through which power can be provided to control the output device.

During normal operation, the processor sends to the logic circuit the output status determined by the logic program. If the output is energized, the signal from the processor is fed to the logic section and passed through the isolation circuit, which will switch the power to the field device.

The switching section generally uses a Triac™ or a silicon controlled rec-

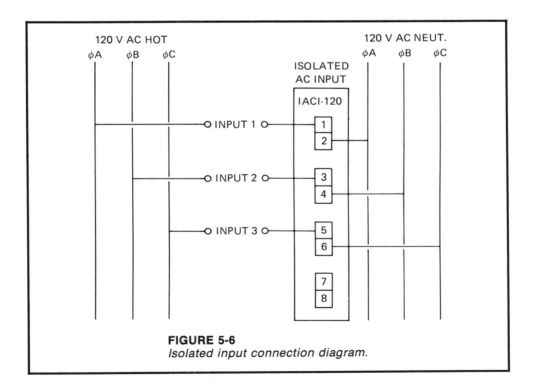

FIGURE 5-6
Isolated input connection diagram.

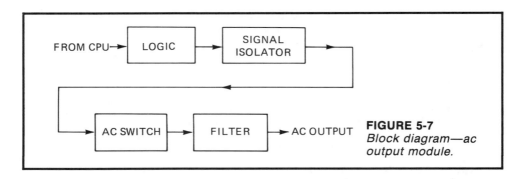

FIGURE 5-7
Block diagram—ac output module.

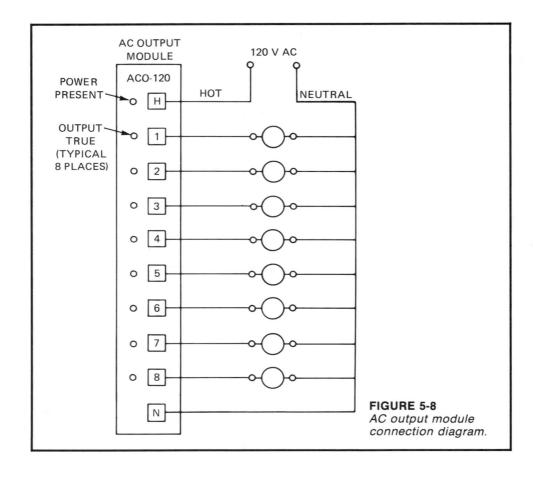

FIGURE 5-8
AC output module connection diagram.

tifier (SCR) to switch the power. The ac switch is normally protected by an RC snubber and often a metal oxide varistor (MOV), which is used to limit the peak voltage to some value below the maximum rating and also to prevent electrical noise from affecting the module operation. A fuse may be provided in the output circuit to prevent excessive current from damaging the ac switch. If the fuse is not provided in the circuit, it should be user-supplied.

As with input modules, the output module may provide light-emitting diode (LED) indicators to show the operating logic. If the circuit contains a fuse, a fuse status indicator may also be incorporated. An ac output module connection diagram is illustrated in Figure 5-8. Note that the switching voltage is field-supplied to the module.

Direct Current (dc) Output Module   The dc output module is used to switch direct current loads. Functional operation of the dc output is similar to the

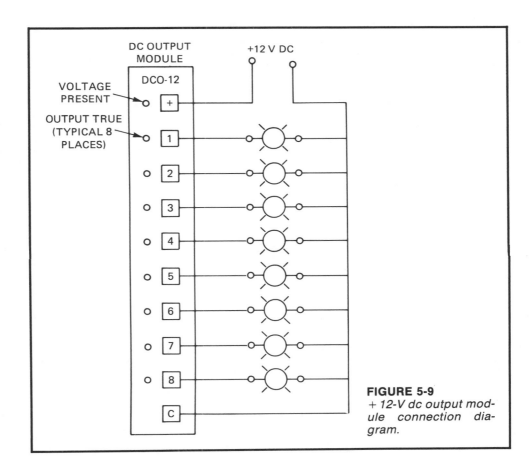

FIGURE 5-9
+ 12-V dc output module connection diagram.

ac output; however, the power circuit generally employs a power transistor to switch the load. Like Triacs, transistors are also susceptible to excessive applied voltages and large surge currents, which could result in overheating and a short circuit. To prevent this condition from occurring, the power transistor will normally be protected. Figure 5-9 shows a +12-V dc output module connection diagram.

Dry Contact Output Module The contact output module allows output devices to be turned ON or OFF by a normally open (NO) or normally closed (NC) set of relay contacts. The advantage of relay or dry contact outputs is that you have electrical isolation between the power output signal and the logic signal. Figure 5-10 shows a dry contact output module with 4 normally open contacts. It also shows that there is complete electrical isolation when the contacts are open, i.e., there is no current flow.

The contact output can be used to switch either ac or dc loads but are normally used in applications such as multiplexing analog signals, switching small currents at low voltage, and interfacing with dc drives for controlling different voltage levels. High power contact outputs are also available for applications that require switching of high currents.

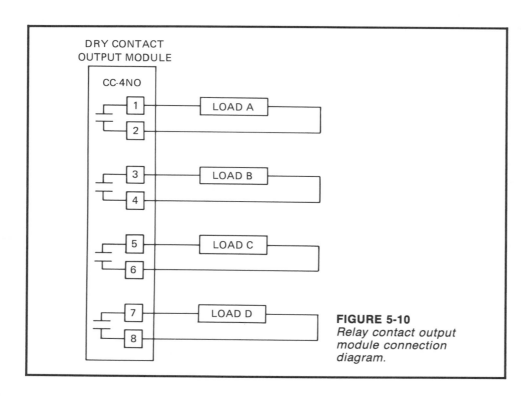

FIGURE 5-10
Relay contact output module connection diagram.

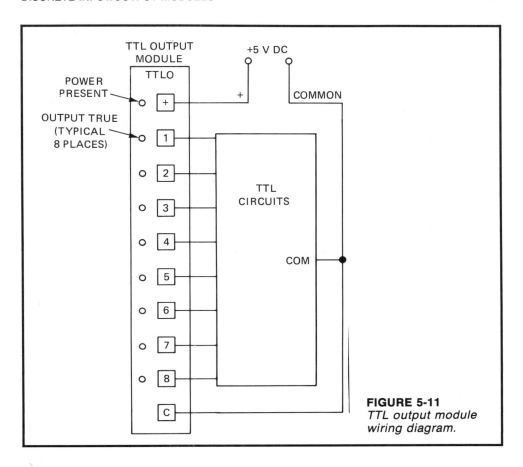

FIGURE 5-11
*TTL output module
wiring diagram.*

TTL Output Module   The TTL output module allows the controller to drive output devices that are TTL compatible, such as seven-segment LED displays, integrated circuits, and various 5-V dc devices. These modules generally require an external +5-V dc power supply with specific current requirements. A typical wiring diagram for a TTL output module is shown in Figure 5-11.

Isolated Output Module   An isolated ac output interface is shown in Figure 5-12. Note, in the isolated module wiring diagram are controlling three loads (A, B, C) that are connected to three different phases of ac power. The advantage of these modules is that we do not have to be greatly concerned with the various ac voltage phases in our process plant. The disadvantages are that we increase the amount of wiring required and decrease the number of available inputs per module by a factor of two (2).

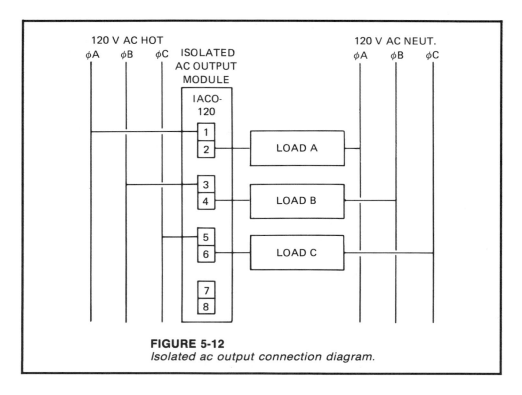

FIGURE 5-12
Isolated ac output connection diagram.

Analog and Digital I/O Modules

The availability of low-cost integrated circuits and the introduction of the microprocessor early in the 1970s greatly increased the capabilities for arithmetic operation and data manipulation in programmable controller systems. This expanded processing capability led to the introduction of sophisticated analog and digital input/output modules.

Analog input modules allow measured quantities to be input from process instruments and other devices that provide analog data, while analog output modules allowed control of devices that require a continuous analog signal.

The digital interfaces are like the discrete I/O modules in that the processed signals are discrete. The difference, however, is that with the discrete I/O only a single bit is required to read an input or control an output. Digital modules allow a group of bits to be interfaced as a unit to accommodate devices that require the bits be handled in parallel form, normally in binary coded decimal (BCD) format or in serial form (i.e., pulse inputs or outputs). The analog I/O will allow monitoring and control of analog voltages and currents, which are compatible with many sensors, motor drives, and process instruments. With the use of digital or analog I/O, most process variables can be measured or controlled with appropriate interfacing. Table 5-3 lists typical

TABLE 5-3
Typical Analog I/O Field Devices

Input Field Devices	Output Field Devices
Flow transmitters	Analog meters
Pressure transmitters	Electric motor drives
Level transducers	Chart recorders
Temperature transducers	Current-to-pressure units
Analytical instruments	Electric valves
Potentiometers	

devices that are interfaced to the controller with analog modules, and Table 5-4 lists some typical digital input and output devices.

Analog I/O interfaces are generally available for several standard unipolar (single polarity) and bipolar (negative and positive polarity) signals. In most cases, a single input or output interface can accommodate two or more different ratings and can satisfy either a current or voltage requirement. The different ratings will be either hardware (circuit board-mounted switches or jumpers) or software selectable. Table 5-5 lists some standard analog I/O ratings.

Analog Input Modules The analog input interface contains the circuitry necessary to accept analog voltage or current signals from field devices. The voltage or current inputs are converted from an analog to a digital value by an analog-to-digital converter (ADC). The conversion value, which is proportional to the analog signal, is passed through to the controller's data bus and stored in a memory location for later use.

Typically, analog input interfaces have a very high input impedance, which allows them to interface field devices without signal loading. The input line from the analog device generally uses shielded conductors. The shielded cable greatly reduces the electrical interferences from outside sources. The input stage of the interface provides filtering and isolation circuits to protect the module from additional field noise.

TABLE 5-4
Typical Digital I/O Field Devices

Input Field Devices	Output Field Devices
Binary encoders	LED displays
Bar code readers	Intelligent panels
Thumbwheel switches	BCD displays

TABLE 5-5
Standard Analog I/O Rating

Analog Signal	Analog Signal
1 to 5 V dc	4 to 20 mA
0 to 5 V dc	0 to 20 mA
−5 to 5 V dc	−20 to +20 mA
−10 to +10 V dc	0 to 10 V dc

Most analog modules are designed to sense up to 16 single-ended or 8 differential analog input signals representing flow, pressure, level, etc. It then converts them to a proportional 12-bit binary or four-digit BCD value. Inputs to a particular module must, in general, be all single-ended or all differential, and the type of signal is either hardware or software selectable. The converted

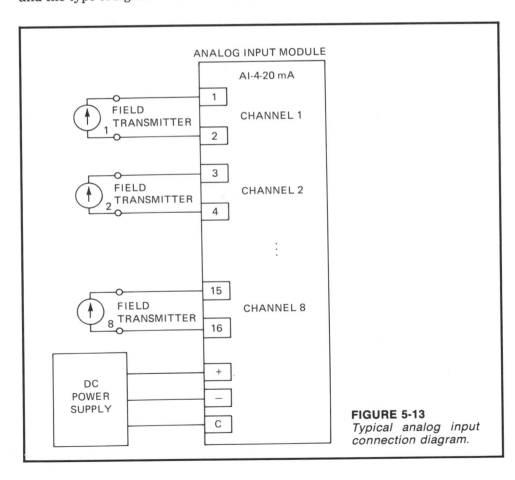

FIGURE 5-13
Typical analog input connection diagram.

signals are stored in memory in the module and are sent to the processor memory in groups or blocks of data.

The control program uses configuration data to set up the analog module. Typical configuration information includes range selection (i.e., 1 to 5 V dc, 4 to 20 mA, etc.) and signal scaling. This configuration data is normally downloaded only once, at the start of a program, using some type of block transfer write instruction.

A typical analog input connection is illustrated in Figure 5-13.

Data is sent to the programmable controller using a block transfer read instruction. A typical sample of data transferred is shown in Figure 5-14. The block transfer instructions are explained in a later chapter.

Analog Output Modules   The analog output modules receive data from the processor, which is translated into a proportional voltage or current to control an analog field device. The digital data is passed through a digital-to-analog converter (DAC) and sent out in analog form. Isolation between the output circuit and the logic circuit is generally provided through optical couplers. These output modules normally require an external power supply with certain current and voltage requirements. Figure 5-15 illustrates a typical device connection for an analog output module.

Binary Coded Decimal Input Modules   The binary coded decimal (BCD) input modules provide parallel communication between the processor and input devices, such as thumbwheel switches. This type of module is generally

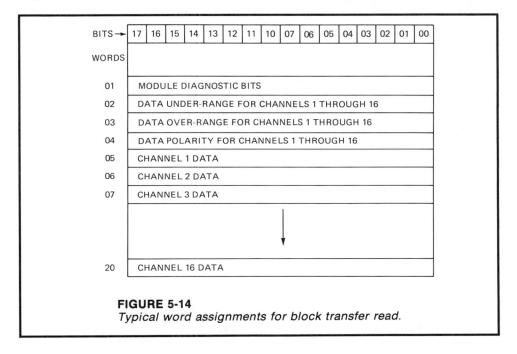

FIGURE 5-14
Typical word assignments for block transfer read.

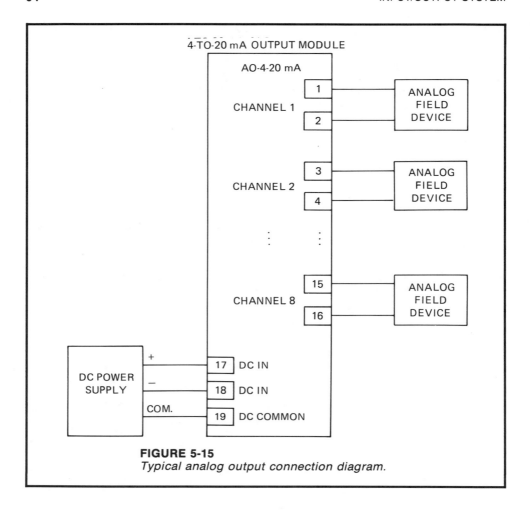

FIGURE 5-15
Typical analog output connection diagram.

used to input parameters into specific data locations in memory to be used by the control program. Typical parameters are timer and counter presets and process control set point values.

These modules generally accept voltages in the range of 5 V dc (TTL) to 24 V dc and are grouped in a module containing 16 or 32 inputs, which correspond to one or two I/O registers. Data manipulation instructions, such as GET or Block Transfer Read, are used to access the data from the register input interface. Figure 5-16 illustrates a typical device connection for a BCD input module. In this module, each bit (1 or 0) from the thumbwheel controls one bit location in a word location in the PC.

BCD Output Modules   This digital module provides parallel communication between the processor and an output device, such as a seven-segment LED display or a BCD display. The BCD output modules can also be used with TTL

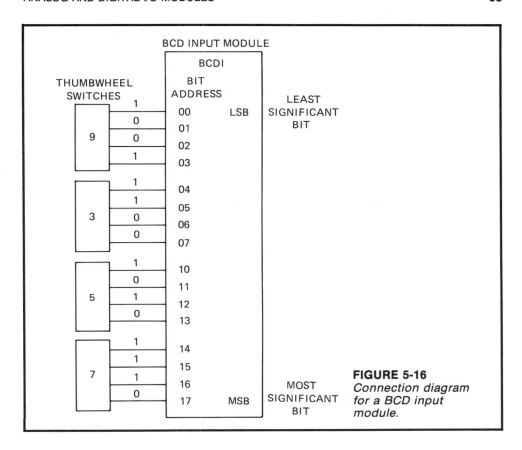

FIGURE 5-16
Connection diagram for a BCD input module.

logic loads that have low current requirements. The BCD output module generally provides voltages that range from 5 V dc (TTL) to 30 V dc and have 16 or 32 output lines (one or two I/O registers).

When information is sent from the processor through a data transfer instruction, the data is latched in the module and made available at the output terminals of the module. Figure 5-17 shows a BCD output module connection.

Encoder/Counter Input Module   The encoder/counter input module provides a high-speed counter, external to the processor, which responds to input pulses sensed at the interface. This counter's operation is normally independent of either program scan or I/O scan. The reason for this is fairly simple: if the counter were dependent upon the PC program, high-speed pulses would be missed during a program scan. Typical applications of the encoder/counter interface are operations that require direct encoder input to a counter, which is capable of providing direct comparison outputs.

The encoder/counter input module accepts input pulses from an incremental encoder, which provides pulses that signify position when the encoder

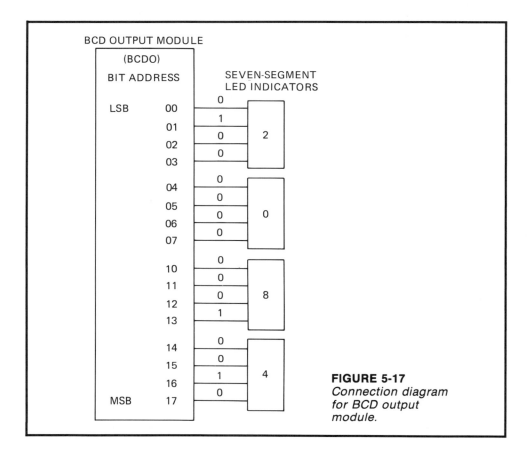

FIGURE 5-17
Connection diagram
for BCD output
module.

rotates. The pulses are counted and sent to the processor. Absolute encoders are generally used with interfaces that receive BCD or Gray code data, which represents the angular position of the mechanical shaft being measured.

During normal operation, the modules receive input pulses, which are counted and compared with a preset value selected by the operator. The counter input module normally has an output signal available, which is energized when the input and preset counts are equal; however, this is not needed in most PCs. Since the data is available in the CPU, the programmer can use a comparison function to drive an output in the control program.

The data communication between the encoder/counter interface and the CPU is bidirectional. The module accepts the preset count value and other control data from the CPU and transmits data and status to the PC memory. The output controls are enabled from the control program, which instructs the module to operate the outputs according to the count values received. The CPU, using the control program, enables and resets the counter operation.

Intelligent I/O Modules

In the previous sections, we discussed discrete, analog, and digital I/O modules that will normally cover 90% of the I/O applications encountered in PC systems. However, to process certain types of signals or data efficiently, microprocessor-based intelligent modules are required. These special interfaces include those that condition input signals, such as thermocouple modules, or other signals that cannot be interfaced using standard I/O modules. These intelligent modules can perform complete processing functions, independent of the CPU and the control program scan. In this section, we will discuss the most commonly available intelligent modules, such as thermocouple input, stepper motor output, and control loop modules.

Thermocouple Input Modules   A thermocouple (T/C) input module is designed to accept inputs directly from a T/C and provides cold junction compensation to correct for changes in cold junction temperatures. The operation of this type of module is similar to the standard analog input with the exception that very low-level signals are accepted from the T/C (approximately 43 mV at maximum temperature for a J-type T/C). These signals are filtered, amplified, and digitized through an A/D converter and then sent to the CPU on command from a program instruction. The data is used by the control program to perform temperature control and/or indication. Figure 5-18 illustrates a typical thermocouple input connection diagram.

Stepping Motor Module   The stepping motor module generates a pulse train that is compatible with stepping motor translators (see Figure 5-19). The pulses sent to the translator normally represent distance, speed, and direction commands to the motor.

The stepping motor interface accepts position commands from the control program. Position is determined by the preset count of output pulses, a forward or reverse direction command, and the acceleration or deceleration command for ramping control is determined by the rate of output pulses. These commands are generally specified during program control, and, once the output interface is initialized by a start, it will output the pulses according to the PC program. Once the motion has started, the output module will generally not accept any commands from the CPU until the move is completed. Some modules may offer an override command that will reset the current position, which must be disabled to continue operation. The module also sends data regarding its status to the PC processor.

A typical stepping motor connection diagram is shown in Figure 5-19.

Control Loop Module   The control loop module is used in closed-loop control where the proportional-integral-derivative (PID) control algorithm is required. Some manufacturers call this interface the PID module, and it is typically applied to any process operation that requires continuous closed-loop control.

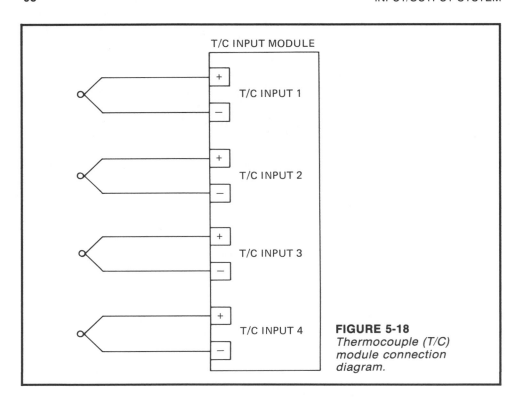

FIGURE 5-18
*Thermocouple (T/C)
module connection
diagram.*

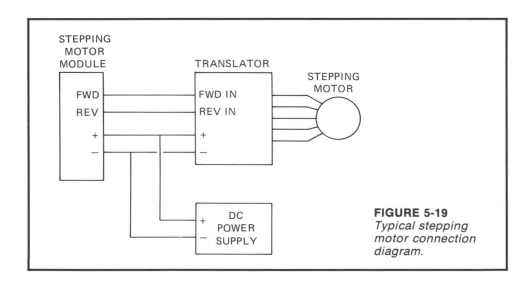

FIGURE 5-19
*Typical stepping
motor connection
diagram.*

The control algorithm implemented by this module is represented by the following equation:

$$V_{out} = K_P e + K_I \int e \, dt + K_D \frac{de}{dt}$$

where:

V_{out} = output control variable
e = PV − SP = error
K_P = the proportional gain
K_I = the integral gain
K_D = the derivative gain

The module receives the process variable (PV), compares it to the set point (SP) selected by the operator and computes the error difference. The operator or control engineer will determine the values of the control parameters (K_P, K_D and K_I), based on the application.

The information sent to the PID module from the PC processor is primarily the control parameters and set points. Depending on the PID module used, data can be sent to describe the update time, which is the period in which the output variable (V_{out}) is updated, and the error dead band, which is a quantity that is compared to the error signal. If the error is less than or equal to the signal error, no update takes place. Some modules provide square root extraction of the process variable, which can be used to obtain a linearized scaled output for use on flow control loops.

Communications I/O Modules

The four common I/O modules used in a programmable controller system to communicate among system components are the ASCII module, local I/O adapter module, the serial data module, and the network interface module.

ASCII Communications Module The ASCII communications module is used to send and receive alphanumeric data between peripheral equipment and the controller. Typical peripheral devices with ASCII I/O include printers, video monitors, digital display instruments, etc. This special I/O module, depending on the manufacturer, is available with a communications circuitry interface that includes onboard memory and a dedicated microprocessor. The information exchange interface generally takes place via an RS-232C, RS-422, or a 20-mA dc current loop communications link.

The ASCII module will generally have its own RAM memory, which can store blocks of data that are to be transmitted. When the input data from the peripheral is received at the module, it is transferred to the PC memory through a data transfer instruction at the PC I/O data bus speed. All the initial communication parameters, such as parity (even or odd) or non-parity, number of stop bits, and communication rate, are hardware selectable or selectable through software. A typical ASCII module is shown in Figure 5-20.

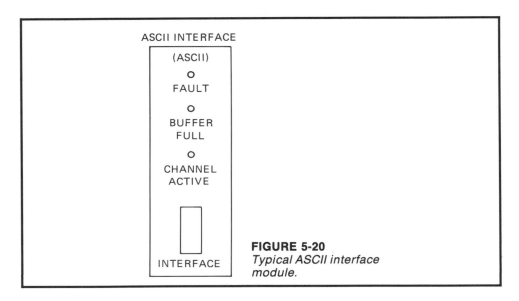

FIGURE 5-20
Typical ASCII interface
module.

Local I/O Adapter Modules   The local I/O adapter modules are used in larger programmable controller systems to allow I/O subsystems to be remotely located from the processor. The remote subsystem is used to interface with a unit process using a standard I/O rack and the required I/O modules. The rack will include a dc power supply to drive the internal circuitry of the I/O modules and a remote I/O adapter module that provides the communications with the processor unit. The I/O capacity of a single subsystem normally ranges from 32 to 256 points.

The subsystems are normally connected to the processor using a serial or star configuration. The distance from the processor to a given remote I/O rack normally ranges from 1000 ft to several miles, depending on the programmable controller type. We will designate a remote I/O adapter module with the letters "ADP". The *serial* configuration is shown in Figure 5-21.

A *star* communications system arrangement is shown in Figure 5-22. It normally requires that the central processing unit have a special circuit card called a scanner module to interface to the remote I/O racks in the system.

Remote I/O arrangements offer large cost savings on wiring materials (wire and conduit) and labor for large control systems in which the field instrumentation is in clusters at several remote process areas. If the processor is located in a main control room or some other central location, only the communication cable needs to be run between the processor and the field I/O racks, instead of hundreds or thousands of field wires.

Remote I/O arrangements also have the advantage of allowing subsystems to be installed and tested independently, as well as allowing maintenance and troubleshooting on individual stations while other units continue to operate.

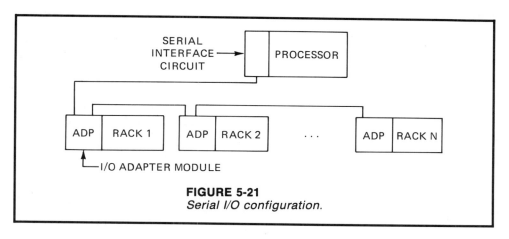

FIGURE 5-21
Serial I/O configuration.

Serial Data Communications Module   The serial data communications module is normally used to communicate between the programmable controller and a computer with a serial output port, such as a personal computer with a serial RS-232 port. A typical application is shown in Figure 5-23, where a personal computer with control program and/or process graphic display software is connected via a serial communications cable and module to a programmable controller. In this configuration, the computer terminal can be used as a ladder programming terminal at a remote location to debug or troubleshoot the control program in the system or to develop or display animated process graphics screens. The data communications module has the model number RS-232 in this treatment.

Network Communications Modules   Network interface modules are designed to allow a number of programmable controllers and other computer-

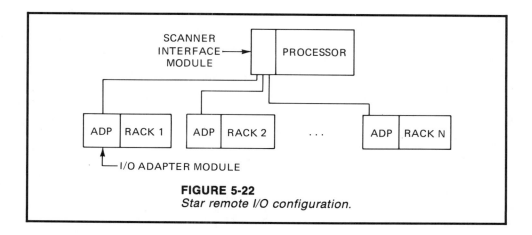

FIGURE 5-22
Star remote I/O configuration.

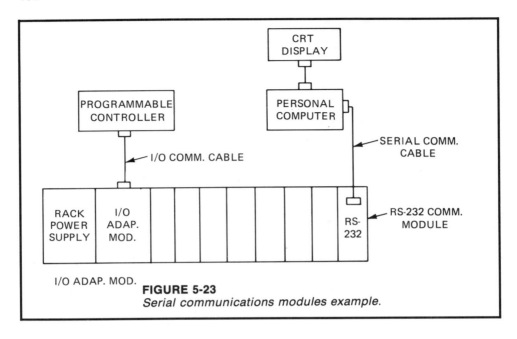

I/O ADAP. MOD.

FIGURE 5-23
Serial communications modules example.

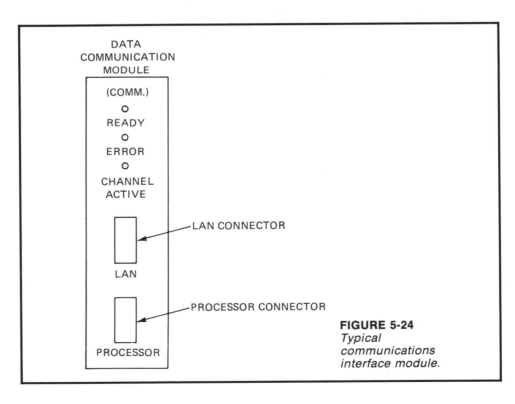

FIGURE 5-24
Typical communications interface module.

based devices to communicate over a local area communication network described in a later chapter. This local area network (LAN) is able to transfer data and control information from one system to another at a fixed data transmission rate. Therefore, the control of an industrial facility can be distributed over a large number of programmable controllers and intelligent devices. In such a system, information is easily exchanged between control systems, but each system can independently control its part of the industrial plant. This greatly improves the reliability of the plant control system, since sections of the plant can be down for modification or maintenance, but the remaining parts of the plant can continue to operate and produce product.

A typical data communication module is shown in Figure 5-24.

Designing I/O Systems To correctly design I/O systems, the programmable controller manufacturer's specifications must be consulted and followed to prevent faulty operation or equipment damage. These specifications place limitations not only on the module but also on the field equipment it operates. The specifications fall into three categories: electrical, mechanical, and environmental.

Electrical Specifications The typical electrical specifications for I/O modules include the following: (1) input voltage rating, (2) input current rating, (3) input threshold voltage, (4) output voltage rating, (5) output current rating, (6) output power rating, and (7) backplane current requirements.

The *input voltage (ac or dc) rating* specification lists the magnitude and type of signal the module will accept. In some cases a range of input voltages instead of a fixed value is stated in the specification. In this case, the maximum and minimum acceptable working voltages for continuous operation are listed. For example, the working voltage for a 120 ac input module might be listed as 95 to 135 V ac.

The *input current rating* defines the minimum input current required at the module's rated voltage that the field device must be capable of supplying to operate the input module circuit.

The *input threshold voltage* is the voltage at which the input signal is recognized as being ON or true. Some input modules also have an OFF voltage value at which the input is OFF or false. For example, the ON voltage for TTL input modules is defined as 2.8 V dc or greater and the OFF voltage is defined as any voltage less than 0.8 V dc.

The *output voltage rating* specifies the magnitude and type of voltage source that can be controlled within a stated tolerance. An output module rated at +24 V dc, for example, might have a working range of 20 to 28 V dc.

The *output current rating* defines the maximum current that a single output circuit in a module can safely carry under load. This current rating is normally specified as a function of the output circuit component's electrical and heat dissipation characteristics at an ambient temperature range (typically 0° to 60°C). As the ambient temperature increases, the output current is normally derated. Exceeding the output current rating can result in a permanent short circuit or other damage to the output module.

The *output power rating* specifies the maximum total power that an output module can dissipate with all outputs energized. The output power rating for a single output is calculated by multiplying the output voltage rating by the output current rating. For example, if a 120-V ac output module has a current rating of 2 amps, the power rating is $P = I \times V$ or $P = 2 A \times 120 V = 240$ watts.

The *backplane current requirement* lists the current demand that a particular I/O module internal circuitry places on the rack power supply. The system designer must add up the current requirements of all the installed modules in an I/O rack and compare the value with the maximum current that can be supplied to determine if the power supplied is correct. If the rack power supply current is too low, intermittent and faulty operation of the system will result.

TABLE 5-6
Typical I/O Module Specifications

Model Number	Module Description	Number of I/O Points	External Voltage Required	Back-Plane Current	I/O Power Rating
AI-4-20 mA	Analog input (4 to 20 mA)	8/16	±15 V dc	400 mA	100 mW
AO-4-20 mA	Analog output (4 to 20 mA)	8/16	±15 V dc	400 mA	100 mW
ACI-120	120 V ac input	8	120 V ac	200 mA	240 W
ACI-220	220 V ac input	8	220 V ac	200 mA	440 W
DCI-12	12 volt dc input	8	12 V dc	200 mA	24 W
DCI-24	24 volt dc input	8	24 V dc	200 mA	48 W
DCI-48	48 volt dc input	8	48 V dc	200 mA	96 W
IACI-120	Isolated ac input	4	None	200 mA	240 W
ACO-120	120 V ac output	8	120 V ac	250 mA	240 W
ACO-220	220 V ac output	8	220 V ac	250 mA	440 W
DCO-12	12 volt dc output	8	12 V dc	250 mA	24 W
DCO-24	24 volt dc output	8	24 V dc	250 mA	48 W
DCO-48	48 volt dc output	8	48 V dc	250 mA	96 W
CC-4NO	4 NO contact output	4	None	500 mA	240 W
TTLI	TTL input	8	5 V dc	150 mA	100 mW
TTLO	TTL output	8	5 V dc	200 mA	100 mW
IACO-120	Isolated ac output	4	None	250 mA	240 W
ASCII	ASCII communication	1	None	500 mA	N/A
ADP	Local I/O adapter	1	None	500 mA	N/A
RS-232	Serial communication	1	None	400 mA	N/A
Comm	Data communication	1	None	250 mA	N/A

Some typical specifications for I/O modules are given in Table 5-6.

An example problem will illustrate a typical I/O system design application.

EXAMPLE 5-1

Problem: Calculate the backplane current requirements for the programmable controller system shown in Figure 5-25. Assume the rack power supply and the backplane power bus are both rated at 5 amps.

Solution: There are three ac input modules, three ac outputs, an I/O adapter interface, and a serial data communications module mounted in the rack, so we use the I/O specification table to find the total current required as follows:

(a) ACI-120:	3×200 mA	= 600 mA
(b) ACO-120:	3×250 mA	= 750 mA
(c) ADP:	1×500 mA	= 500 mA
(d) RS-232:	1×400 mA	= 400 mA
Total Current Required		= 2250 mA

Therefore, the total backplane current demand is 2250 mA or 2.5 amps. This is less than the rack power supply and backplane current rating of 5 amps, so I/O system current demand is acceptable.

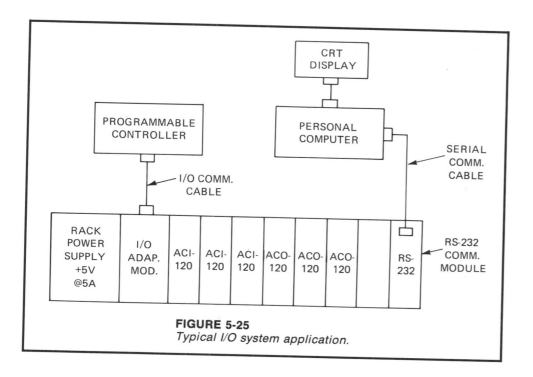

FIGURE 5-25
Typical I/O system application.

Mechanical and Environmental Specifications   The typical mechanical specifications are I/O points per module and wire size. The I/O points per module simply defines the number of field points that are controlled or sensed by a module. Typically, modules will have 1, 2, 4, 8, 16, or 32 points per module. The higher density modules require higher operating current, so the backplane current must be carefully checked. The number of wires is also increased, and it might be a problem for larger gage wires. The wire size specification defines the number of conductors and the largest wire gage that the I/O terminal points will accept. For example, a typical wire specification for an ac input/output module is 2-14 AWG wires per terminal.

The important environmental specifications are the ambient temperature rating and the humidity rating. The ambient temperature specification defines the maximum temperature of the surrounding air in which the I/O system will operate properly. This rating is based on the heat dissipation characteristics of the circuit components inside the I/O modules, which are much higher than the module ambient temperature rating. A typical value of ambient temperature rating is 0° to 60°F. Exceeding the ambient temperature can be dangerous because the internal circuits of the modules will act erratically, resulting in undesirable outputs to the process being controlled.

The humidity specification is typically 5% to 95% without condensation. The system designer must insure that the humidity is properly controlled in the plant area where the I/O system is installed.

Adherence to the PC manufacturer's specifications will insure proper and safe operation of the control system.

EXERCISES

5.1 Describe the main functions and advantages of programmable controller I/O equipment racks.

5.2 What are the purposes and functions of the backplane on a typical I/O equipment rack?

5.3 Describe the operation and purposes of the five circuits in a typical ac input module.

5.4 Draw a wiring diagram for a typical +24-V dc input module with four temperature high switches (THS-100, 101, 102, and 103) and three pressure low switches (PLS-210, 211 and 212) connected to the module.

5.5 Describe the operation of the internal circuits in a typical ac output module.

5.6 Draw the wiring diagram for a BCD output module connected to a four-digit LED display. Assume the output of the module is the decimal number 2783 and show the binary bits out to each LED.

5.7 Calculate the backplane current requirements for a 16-slot I/O rack with the following modules installed: (a) four 12-V dc input modules, (b) six 12-V dc output modules, (c) four analog input modules, and (d) two analog output modules. Assume the rack power supply and the backplane circuit card are rated at 5 amps.

BIBLIOGRAPHY

1. Jones, C. T., and Bryan, L. A., *Programmable Controllers, Concepts and Applications,* International Programmable Controls, Inc., 1983.
2. Hughes, T. A., *Basics of Measurement and Control,* ISA Publications, 1988.
3. *Processor Manual PLC-5 Family Programmable Controllers,* Allen-Bradley Co., Inc., 1987.
4. *Programming and Operations Manual, PLC-2/30 Programmable Controllers,* Allen-Bradley Co., Inc., 1988.
5. *Allen-Bradley Industrial Computer and Communications Group Product Guide,* Allen-Bradley Co., Inc., 1987.
6. *Allen-Bradley Programmable Controller Products,* Allen-Bradley Co., Inc., 1987.

6

Memory and Storage

Introduction   Programmable controller systems store information and programs by either electronic or mechanical means. Electronic methods, generally called memory, are used to store the control program for the programmable controller system, and it is usually located in the same housing as the processor. The information stored in memory determines how the input and output data will be processed by the programmable controller.

However, electronic memory circuits are expensive and the most common varieties lose data when power is interrupted for even a split second, so there must be mechanical means to permanently back up control data and instructions. Commonly called storage, these mechanical devices once stored data in the form of holes punched in paper tape or cards; now the holes have been supplanted by infinitesimally small areas of magnetism on recording tape or disks.

Storage devices for programmable controllers are mechanical in the sense that retrieving magnetically stored data requires the use of machinery such as tape player/recorders or disk units in order to locate the data and convert it into electronic pulses acceptable to the programmable controller. Storage devices are much slower than electronic memory, but their main advantage is permanent storage. For example, information or programs recorded on magnetic tape or disk can be retained for years.

Memory   Programmable controller memories can be visualized as a two-dimensional array of storage cells, each of which can store a single bit of information in the form of a 1 or a 0. This single "bit" gets its name from BInary digiT. A bit is the smallest structural unit of memory and stores information in the form of 1s and 0s. Ones and zeros are not actually in each cell; each cell has a voltage present, indicated by a one, or not present, indicated by a zero.

The bit is set or ON if the stored information is 1, and OFF if the stored information is 0. In most cases it is necessary for the processor to handle more than a single bit. For example, when transferring data to and from memory, storing numbers, and programming codes, a group of bits called a byte or word is required. A byte is defined as the smallest group of bits that can be handled by the CPU at one time. In programmable controllers, byte size is normally 8 bits but can be smaller or larger depending on the specific computer being used.

Memory capacity is specified in thousands or "K" increments where 1K is 1024 words (i.e., $2^{10} = 1024$) of storage space in most cases. Programmable controller memory capacity may vary from less than one thousand bits to over 64,000 words (64K words), depending on the programmable controller manufacturer. The complexity of the control plan will determine the amount of memory required.

Word length is usually 1 byte or more in length. For example, a 16-bit word consists of 2 bytes. Typical word lengths in programmable controllers are 8 or 16. A 16-bit word is shown in Figure 6-1.

Most programmable controllers use the octal numbering system to identify each bit in a given word as shown in the figure. The most significant bit (MSB) is bit 17, and the least significant bit (LSB) is bit 00.

A simple 64-cell memory array is shown in Figure 6-2. This array consists of 8 rows and 8 columns. This 64-bit array requires only 6 bits to address a given cell. A cell is normally an electronic circuit called a flip-flop that can have a value of 5 volts (logic 1) or 0 volts (logic 0). To retrieve data from the memory array, row and column address decoders select the appropriate cell.

These memory arrays are normally provided by integrated circuits (ICs). A typical IC unit contains many thousands of memory cells arranged in various ways. A 8K-bit (8096 cells) memory integrated circuit, for example, could be arranged as 8K memory cells of 1 bit each, or 1K bytes of 8 bits each. The number of groups (bits, bytes, or words) addressed is a function of 2^n, for example $1K = 2^{10}$, $4K = 2^{12}$, $8K = 2^{13}$, etc. The value n is the number of address bits needed to select each separate group. For 1000 words it is necessary to use 10 bits to address each word of storage, the word size being 4 bits, 8 bits, 16 bits, etc. For a 1K × 8 memory the IC would require 10 address bits to select 1K words of memory. A typical 1K × 8 memory chip is shown in Figure 6-3. The IC has 8 pins for input and output data bits, 10 pins for selecting

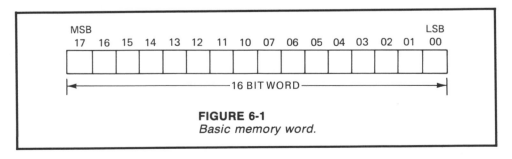

FIGURE 6-1
Basic memory word.

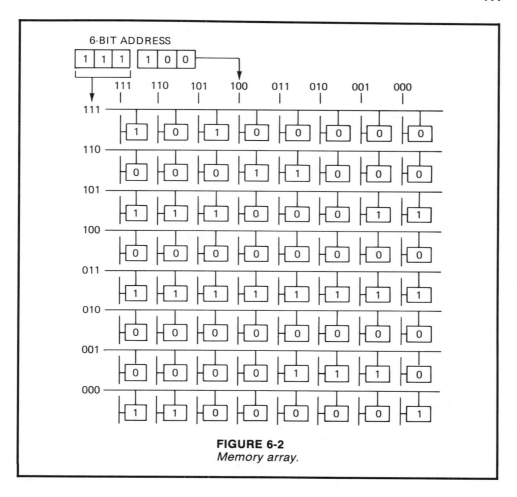

FIGURE 6-2
Memory array.

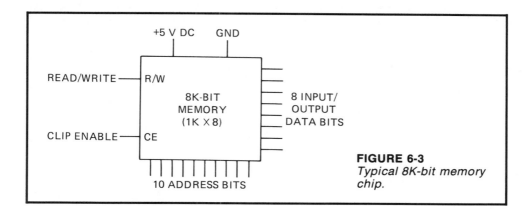

FIGURE 6-3
Typical 8K-bit memory chip.

addresses, 2 pins for control signals (chip enable and read/write), and 2 pins for dc power. The two power supply pins are used for +5 V dc and ground. The Read/Write (R/W) control signal is used to determine whether the data bits are read into memory (R/W signal low) or data is sent out from memory (R/W is high). The chip enable control signal is used to select operation of each separate chip when a group of integrated circuits is used to provide a larger memory than is provided by only one chip.

Memory Types
This section will discuss the types of memory generally used in programmable controllers and their applications to the type of program and data stored. In selecting the type of memory to be used, a system designer is concerned with volatility and ease of programming. He is concerned with volatility because memory holds the process control program, and, if this program is lost, production in a plant will be down. Ease in altering the memory is important since the memory is involved in any interaction that takes place between the user and the processor. This interaction begins with the initial system programming and debugging and continues with on-line changes, such as changing timer and counter preset values.

Random Access Memory (RAM)
Random access memory is designed so that data or information can be written into or read from any unique location. Figure 6-4 shows that data can be placed into RAM memory using the *write* mode and data can be retrieved from RAM using the *read* mode. The *address* input to the RAM specifies the location or the address of the data to be read or the location to be written into.

Programmable controllers, for the most part, use RAM with battery backup for application memory. RAM provides an excellent means for easily creating and altering a control program as well as allowing data entry. In comparison to some other memory types, RAM is relatively fast. The only important disadvantage of battery-supported RAM is that it requires a battery that might fail at a critical time, but it is sufficient for most programmable controller applications.

RAM memory is an integrated circuit chip that stores individual bits of data in multiple rows and columns of cells. This row and column arrangement

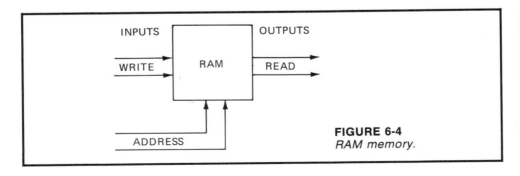

FIGURE 6-4
RAM memory.

allows each cell to have a unique designation, called an address. This address consists of a row identifier and a column identifier, both expressed as binary numbers as discussed earlier.

Read-Only Memory (ROM)　Read-only memory is designed to permanently store a fixed program, which normally cannot or will not be changed. It gets its name from the fact that its contents can be read but not written into or altered once the data or program has been stored. Figure 6-5 shows that data can be used only in the *read* mode. As with RAM memory, ROM has an address input to specify the location of the data to be read. Because of their design, ROMs are generally immune to changes due to electrical noise or loss of power. The executive or operating system program of PC is normally stored in ROM.

Programmable controllers rarely use ROM for the control applications program memory. However, in applications that require fixed data, ROM offers advantages where speed, cost, and reliability are factors. Generally, ROM-based PC programs are produced at the factory by the equipment manufacturer. Once the original set of instructions is programmed, it can never be altered by the user. The manufacturer will write and debug the program using a read/write-based controller or computer and then the final program is entered into ROM. ROM application memory is typically found in dedicated programmable controllers.

Programmable Read-Only Memory (PROM)　The PROM is a special type of ROM that is rarely used in most programmable controller applications. However, when it is used, it will most likely be a permanent storage to some type of RAM. Although PROM is programmable and, like other ROM, has the advantage of nonvolatility, it has the disadvantages of requiring special programming equipment, and, once programmed, it cannot be erased or altered. Any program change would require a new set of PROM chips. PROM memory might be suitable for storing a program that has been thoroughly checked while stored in RAM and will not require further changes or on-line data entry.

Erasable Programmable Read-Only Memory (EPROM)　The EPROM is a special type of PROM that can be reprogrammed after being completely erased

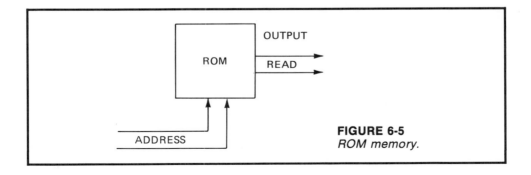

FIGURE 6-5
ROM memory.

using an ultraviolet (UV) light source. The integrated circuit chip for EPROM is built with a window on the top of the chip so the internal memory circuits can be exposed to the UV light. The EPROM can be considered a temporary storage device in that it stores a program until it is ready to be changed. EPROM provides an excellent storage medium for a control program where nonvolatility is required but program changes are not required. Many manufacturers of equipment with built-in PCs use EPROM-type memories to provide permanent storage of the machine program after it has been developed, debugged, and is fully operational.

A control program composed of EPROM alone would be unsuitable if online changes and/or data entries are a requirement. However, many PCs offer EPROM control program memory as an optional backup to battery supported RAM. EPROM, with its permanent storage capability combined with the easily altered RAM, makes a suitable memory system.

Core   Core memory is a read/write memory that gets its name from the fact that it stores individual bits by magnetizing a small ferrite core in the 1 or 0 direction through a write-current pulse. Each core represents a bit and can hold a 1 or 0 even if power is lost. Core memory units are nonvolatile because power is not necessary to keep the core magnetized. The state of each core is also electrically alterable, which makes it a read/write memory unit.

Core memory uses the intimate relationship between electricity and magnetism to provide storage of data. An electric current in a conductor creates a magnetic field around it. The orientation of the magnetism is determined by the current flow, according to the right-hand rule: with the thumb pointed in the direction of current, the fingers of the right hand curl in the direction of the magnetic field. This field can be used to permanently magnetize a doughnut-shaped core of ferromagnetic material as shown in Figure 6-6. Magnetizing the core in the counterclockwise direction, as shown in Figure 6-6, stores a logical one. Reversing the current magnetizes the core in the opposite direction, storing a logical zero.

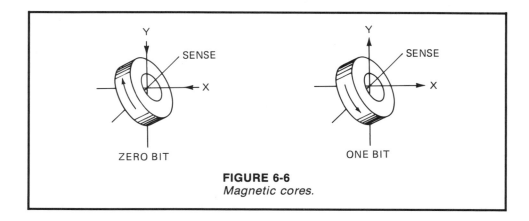

FIGURE 6-6
Magnetic cores.

To permanently magnetize a core, the current must exceed the magnetic threshold of the ferromagnetic material in the core. However, the current does not need to be supplied by a single wire. For example, two wires can each furnish half the threshold current. Furthermore, the wires do not have to be parallel to each other as long as the current that flows in both wires is in a direction that magnetizes the core in a single direction. In fact, the two wires can cross at right angles to each other as shown in Figure 6-6. This method of construction leads to the practical two-dimensional array shown in Figure 6-7. This array consists of horizontal and vertical wires with tiny rings of ferromagnetic material hanging at each intersection of the two wires.

Any one of the doughnuts in the array can be selected individually to write a bit simply by energizing the wires that pass through it. Because the cores can be activated in any sequence imaginable, core memory is termed random access memory or RAM.

To read the memory to determine whether a core contains a 1 or a 0, the appropriate wires are energized with the currents necessary to magnetize the core as a zero. This write operation generates a voltage pulse in the sense wire.

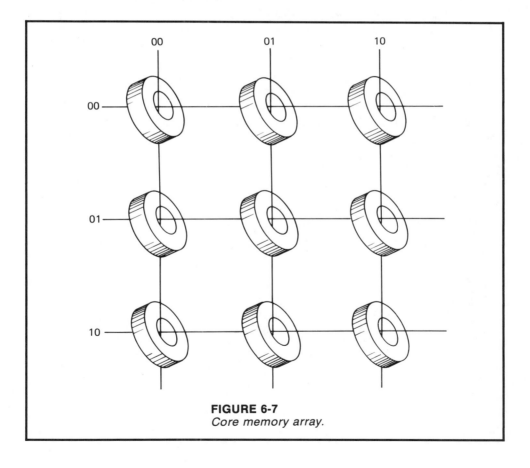

FIGURE 6-7
Core memory array.

The sense line is wired diagonally through the memory array. If a pulse is detected, the computer knows that the content of the core was logic one. It then reverses the two currents through the wires to erase the zero and restore the one. When the same operation is used to read a core containing a zero, no voltage pulse is produced, and the computer interprets the absence of a pulse as a logical zero.

Magnetic core was used in many of the first programmable controllers and is still used in a few today. It provides excellent permanent storage and is easily changed. Some of the disadvantages of using core memory in programmable controllers are its slow speed, high current requirements, relatively expensive cost, and larger physical space requirements.

Storage Units

While RAMs, ROMs, and magnetic cores are used for programmable controller internal memory, other devices such as floppy diskettes (disks) or hard disks are used for external programs and data storage. Information is stored on a disk much as music or video is recorded on tape by magnetizing small areas on the disk in a predefined direction. One direction represents a 1 bit, and the opposite direction would represent a 0 bit in a microcomputer-based system, such as a programmable controller or a personal computer.

Storage on Disks   A typical $5\frac{1}{4}$ inch floppy disk consists of two parts: a disk of thin plastic coated with magnetic material and a protective plastic jacket. A typical floppy disk is shown in Figure 6-8. The magnetic coating is visible through the opening in the jacket. The hole in the center of the disk goes around the drive motor, which spins the disk so that data can be written or read.

If you cover the write-protect notch with a piece of tape, no data on the disk can be changed. Data is stored on a disk in narrow concentric circles called tracks; there are 40 tracks on a standard disk and 80 tracks on the high-

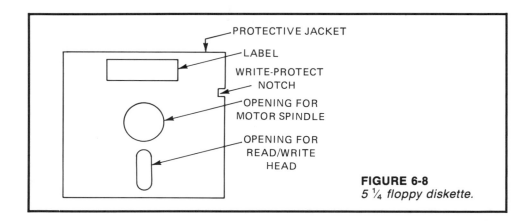

FIGURE 6-8
5 ¼ floppy diskette.

capacity disk used on the IBM personal computer AT℠. A track is divided into smaller areas called sectors, each of which can hold 512 bytes of data (1/2K).

Tracks on a standard disk are numbered 0 through 39 (i.e., 40 tracks); sectors are numbered 1 through 9, for a total of 360 sectors (9 sectors per track times 40 tracks) on each side. Most personal computers have double-sided disk drives, which use both sides of a disk. A double-sided disk can store 368,640 bytes or 360K bytes.

The high-capacity disks on the IBM personal computer AT have 80 tracks, each of which has 15 sectors. A sector still holds 512 bytes, so a high-capacity disk can store 1,228,800 bytes or 1.2 megabytes.

Storage on Tape Information is stored magnetically on tape in straight line tracks. On a tape, related data bytes are grouped into what is termed a record and stored together in one or more blocks along the tape. Because the read/write head cannot jump from place to place on the tape, locating a needed piece of information may require scanning hundreds of inches of tape. In some cases a read operation can take over thirty seconds to complete. Since this time period is normally unacceptable in process or machine control operations, tape units are normally only used to back up programmable controller programs or to archive process data.

The most common tape format is nine track. In this format, the eight bits of each byte of data are recorded simultaneously by eight heads across separate tracks in a line perpendicular to the edge of the tape. The remaining track is used for the parity bit, which is recorded by a separate head.

Floppy Disk Drives A magnetic disk is useless without a disk drive to read or write data on its magnetic surface. Although the drives for floppy disks differ in detail, they have quite a bit in common. For example, they all have read/write heads that transfer data to and from the floppy disks. To store and read data, the read/write heads of all floppy disk drives touch the disk's magnetic surface. Despite the presence of liners to clean the disk, the constant rubbing of the head against the disk can clog the narrow recording gap in the head. This constant abrasion ultimately wears out the floppy disk and head, eventually leaving the disk unreadable or the drive defective.

Memory Organization
The programs and data stored in the memory of programmable controller systems are generally described using four terms: (1) Executive, (2) Scratch Pad, (3) Control Program, and (4) Data Table. The *Executive* is a permanently stored collection of programs that is the operating system for the PC. This operating system directs activities such as execution of the control program, communication with the peripheral devices, and other system housekeeping functions. The *Scratch Pad* is a temporary storage area used by the CPU to store a relatively small amount of data for interim control or calculations. Data that is needed quickly is stored in this area to increase the speed of data manipulation. The *Control Program* area

provides storage for any programmed instructions entered by the user. The *Data Table* stores any data associated with the control program, such as timer/counter preset values and any other stored constants or variables that are used by the main program or the CPU. It also retains the input/output status information.

The storage and retrieval requirements are not the same for the executive, the scratch pad, the control program, and the data table; therefore, they are not always stored in the same types of memory. For example, the executive requires a memory that permanently stores its contents and cannot be deliberately or accidentally altered by loss of electrical power or by the user; therefore, some type of ROM would be used. On the other hand, the user would need to alter the control program and/or the data table for any given application, so some type of RAM unit would be used.

Memory Size

The size of memory is an important factor in designing programmable controller-based control systems. Specifying the correct memory size can save hardware costs and avoid lost time later. Proper calculation of memory size means avoiding the possibility of purchasing a PC that does not have adequate capacity or that is not expandable.

Memory size is normally expandable to some maximum point in most controllers, but it is not expandable in some of the smaller PCs. (Smaller PCs are defined in this context as units that control 10 to 64 input/output devices.) Programmable controllers that handle 64 or more I/O devices are usually expandable in increments of 1K, 2K, 4K, etc. (K represents 1024 or 2^{10} word locations in memory). The maximum memory in the larger controllers is generally 64K.

The stated memory size of a PC is only a rough indication of the memory space available to the user, since some of the memory is used by the controllers for internal functions. Another problem is that there are generally two different word sizes, 8 bit and 16 bit, used by PC manufacturers. Figure 6-9 shows a comparison of 2K of memory based on 8-bit and 16-bit words. It's clear that the 16-bit memory configuration has twice the storage capacity.

The main problem with determining memory size for an application is that the complexity of the control program is not determined until after the equipment is purchased. However, we generally know the number of I/O points in the system before the hardware is procured. Memory size can be estimated from this fact; a good rule is to multiply the number of I/O points by 10 words of memory. For example, if the system has 100 I/O points, the program will generally be equal to or less than 1000 words. We should keep in mind that program size is affected by the sophistication of the control program. If the application requires data handling or complex control algorithms, such as PID control, then additional memory will be required.

After the system designer determines the minimum memory required for an application, he will normally add an additional 25% to 50% more for program changes, modification, or future expansion.

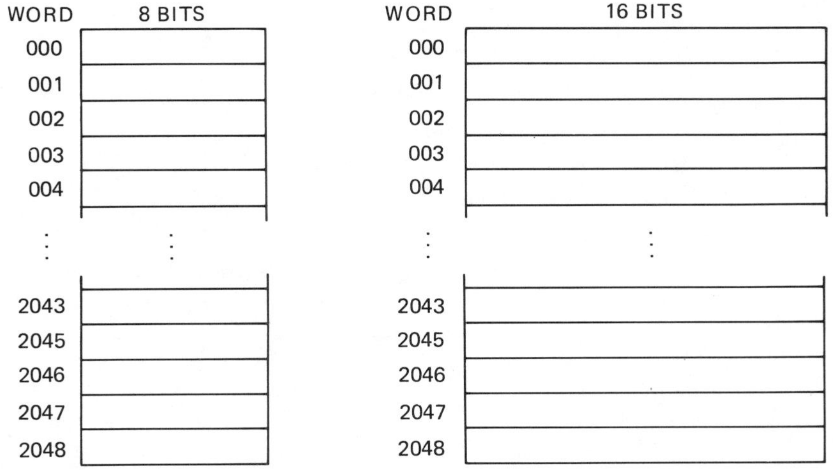

WORD   8 BITS   WORD   16 BITS

000		000	
001		001	
002		002	
003		003	
004		004	

⋮   ⋮   ⋮   ⋮

2043		2043	
2045		2045	
2046		2046	
2047		2047	
2048		2048	

FIGURE 6-9
Comparison of 2K of memory for 8- and 16-bit words.

EXAMPLE 6-1

Problem: Determine memory size needed for a programmable controller system with 173 input points and 125 outputs (assume 25% spare memory capacity).

Solution: Total I/O points = 173 + 125 = 298 points
Memory required = 298 points × 10 words/point + 25%
= (2980 + 2980 × 0.25) words
= 4023 words ≅ 4 K

To accurately size memory, we need to understand the overall organization of PC memory.

Memory Map

The arrangement of programmable controller memory is known as a *memory map*. A memory map is used to show the location of both the system memory and the application memory; it is an aid to the programmable controller programmer in writing the control program. A typical memory map is shown in Figure 6-10 that consists of the following units: executive, processor work area, the input/output table, data table, and the user program.

Normally the programmer does not have access to the executive or the processor work area. These two areas are called the "system memory". Care must be taken not to include this area as part of the available programming

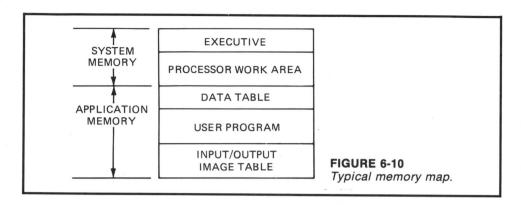

FIGURE 6-10
Typical memory map.

area when sizing the PC memory for an application. On the other hand, the data table and the user program areas are used by the programmer and are called the "application memory". The memory, therefore, consists of three main parts: the system memory, the application memory, and the input/output table.

It is important to note again that some controller manufacturers include system memory in their total specified memory size. A controller with 64K of memory may have system memory of 32K and 2K of I/O image table, leaving only 30K for application memory. Fortunately, most PC vendors will exclude the system memory when they specify memory size, so only the application memory size need be considered when designing a PC control system.

Application Memory   The application memory stores control program instructions and I/O data points that are used by the CPU to perform its control functions. A typical map of the application memory is shown in Figure 6-11. The application memory can be divided into two major areas: the *data table* and the *user program*. All data is stored in the data table, while programming instructions are stored in the user program area.

The data table can be functionally divided into four areas: (1) data storage

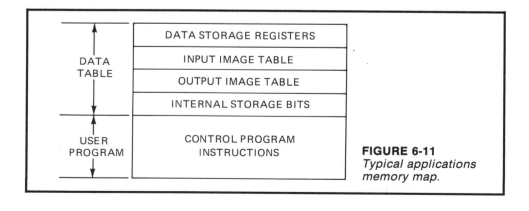

FIGURE 6-11
*Typical applications
memory map.*

registers, (2) input image table, (3) output image table, and (4) internal storage bits.

Data Storage Registers

The information stored in the data storage register is generally process or control information. In general, the three types of storage registers are input registers, holding registers, and output registers. The total number of registers varies depending on the controller memory size and how the data table is configured (ratio of data storage area to program storage area). Values stored in the storage registers are in binary or BCD format. Each register can generally be loaded, altered, or displayed by using the programming unit or a special data entry device offered by most manufacturers.

Input registers are used to store numerical data received, via input interfaces, from devices such as thumb wheel switches, shaft encoders, and other devices that provide BCD input. Analog signals also provide numerical data that must be stored in input registers. The current or voltage signal generated by various analog transmitters is converted by the analog interface. From these analog values, binary representations are obtained and stored in the designated input register. The value contained in the input register is determined by the input device and, therefore, is not alterable from within the controller or via any other form of data entry.

The holding registers are those required to store variable values that are program-generated by instructions (e.g., math, timer, or counter) or constant values that are entered via the programming units or some other data entry method.

Output registers are used to provide storage for numerical or analog values that control various output devices. Typical devices that receive data from output registers are alphanumeric LED displays, recorder charts, analog meters, speed controllers, and control valves. Output registers are essentially holding registers that are designated "output" because of their particular nature (i.e., controlling outputs).

In addition to the I/O tables and storage register areas, some controllers allocate a portion of the data table for storing decimal, ASCII, and binary data. Typically, this table area is used for recipe data, report generation messages, or other data that can be stored or retrieved during program execution.

Input Image Table

The input image table is an array of bits that store the status of digital inputs from the process, which are connected to input modules. The number of bits in the table is equal to the maximum number of inputs. A controller with a maximum of 64 inputs would require an input table of 64 bits. Each connected input has a bit in the input table that corresponds exactly to the terminal to which the input is connected. If the input is ON, its corresponding bit in the table is ON (1). If the input is OFF, the corresponding bit is cleared or turned OFF (0).

The input table is continuously being changed to reflect the current status of the connected input devices. This status information is also being used by the control program.

The input table shown in Figure 6-12 uses a common numbering system that stores field input data starting at word 110 and ending at word 177. In this system the leftmost digit, the number 1, indicates an input to the programmer, the second digit gives the I/O rack number (racks 1 through 7), and the final digit gives the module group number in the I/O rack.

Figure 6-13 shows a typical example of a single input bit in an input image table. In this example, input point 11714 is identified on the memory map.

Output Image Table

The output table is an array of bits that controls the status of digital output devices, which are connected to output interface circuits. The number of bits in the output table is equal to the maximum number of outputs. A controller with a maximum of 512 outputs would require an output table of 512 bits.

The output image table shown in Figure 6-14 uses the same numbering system as the input image table, except the first (the leftmost digit) is the number 0 to signify an output, the second digit gives the I/O rack number (racks 1 through 7), and the final digit gives the module group number in the I/O rack.

Each connected output has a bit in the output table that corresponds exactly to the terminal to which the output is connected. Note that the bits in the output table are controlled by the CPU as it interprets the control program and are updated accordingly during the I/O scan. If a bit is turned ON (1), then the connected output is switched ON. If a bit is cleared or turned OFF (0), the output is switched OFF.

Figure 6-15 shows a typical example of a single output bit in an output image table. In this example, output 01716 is shown on the memory map.

Internal Storage Bits

Most controllers assign an area for internal storage bits. These storage bits are also called internal outputs, internal coils, or internal control bits. The internal output operates just as any output that is controlled by programmed logic; however, the output is used strictly for internal logic programming and

RACK 1	110-117
RACK 2	120-127
RACK 3	130-137
RACK 4	140-147
RACK 5	150-157
RACK 6	160-167
RACK 7	170-177

FIGURE 6-12
Typical input image table.

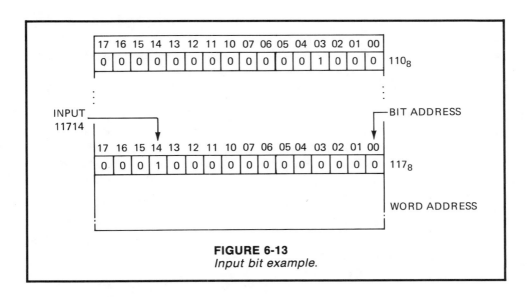

FIGURE 6-13
Input bit example.

RACK 1	010-017
RACK 2	020-027
RACK 3	030-037
RACK 4	040-047
RACK 5	050-057
RACK 6	060-067
RACK 7	070-077

FIGURE 6-14
Typical output image table.

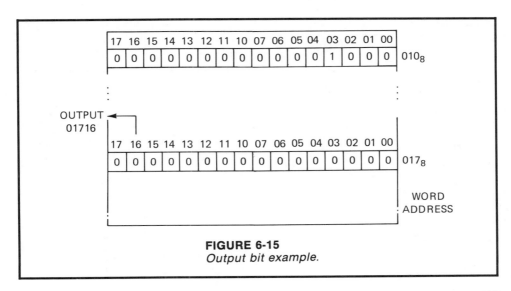

FIGURE 6-15
Output bit example.

does not directly control an output to the process. Internal outputs are used for interlocking logic purposes in the control program.

Internal outputs include the "done" bits on counters and timers as well as internal logic relays of various types. Each internal output bit, referenced by an address in the control program, has a storage bit of the same address. When the control logic is TRUE, the internal (output) storage bit turns ON.

The user program memory area of the application memory is used to store the process control logic program. All the controller instructions that control the machine or process are stored here. The addresses of the real and internal I/O bits are specified in this section of memory. When the PC is in the RUN mode and the control program is executed, the CPU interprets these memory locations and controls the bits in the data table, which corresponds to a real or internal I/O bit. The interpretation of the control program is accomplished by the processor's execution of the executive program.

The maximum amount of user program memory available is normally a function of the controller memory size. In medium and large programmable controllers, the user program size is normally flexible, through altering the data table size so that it meets the minimum data storage requirements. In small PCs, however, the user program size is normally fixed.

Hardware-to-Program Interface
Probably the most important thing to understand about programmable controllers is how process data, sensed by the input modules, is used by the processor to activate output devices to control the process. This hardware-to-program interface occurs in the input/output image tables and was briefly discussed earlier. The instruction address is what connects the software to the hardware. We will present one common method used in programmable controller systems and use this I/O numbering scheme in all examples and applications.

Instruction Address   The addressing method selected uses 5 digits to reference both an I/O image table address and a hardware location. In this system,

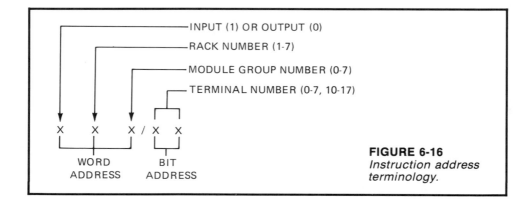

FIGURE 6-16
Instruction address terminology.

the leftmost bit is a 1 for an input and a 0 for an output. The next bit is the rack number, and we have assumed that our system has only 7 racks. The next bit or middle bit is the module group number. The remaining two digits represent the bit address in the I/O image table word and the terminal number in the I/O module (see Figure 6-16).

EXAMPLE 6-2

Problem: List the memory bit address for an input to rack 3, module group 4, terminal 10.

Solution:

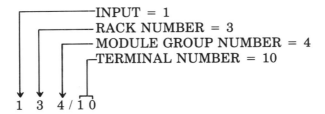

```
                    ┌──────────INPUT = 1
               ┌────────────RACK NUMBER = 3
          ┌─────────MODULE GROUP NUMBER = 4
          ┌─TERMINAL NUMBER = 10


     ↓    ↓   ↓   │
     1    3   4 / 1 0
```

Hardware-to-Software Interface

The hardware-to-software interface is illustrated in Figure 6-17 by showing the operational relationship between the field devices, the input/output image table, and the user program.

In the example shown, when an input device connected to terminal 111/14 is activated, the input module sends a logic 1 to input image table bit 111/14. During the program scan, the processor examines bit 111/14 for an ON or 1 condition. If the bit is ON, the ladder program examine on (⊣ ⊢) is logically TRUE, and a path of logic continuity is established and causes the rung to be TRUE. The processor then sets output image table bit 011/03 to ON (1). The processor will turn ON terminal 011/03 of the output module during the next I/O scan, and the field device wired to this terminal will be energized.

When the input device wired to terminal 111/14 opens, the input module senses the open state, and the input module sets the bit to zero (0) or OFF. The OFF condition is reflected in input image table bit 111/14. During the program scan, the processor examines the bit for an ON (1) condition, and, since the bit is OFF (0), logic continuity is not established and the ladder rung is FALSE. The processor then sets output table bit 011/03 to OFF (0). In the next I/O scan, it instructs the output module to turn off output 011/03, and the output device wired to this terminal is turned OFF.

Data Table Documentation

The documentation of the data table is probably one of the most important functions performed by a system designer. Complete and systematic documentation avoids improper use of data table areas during programming. For example, a programmer might accidentally assign a timer accumulated word to an input location, which would cause

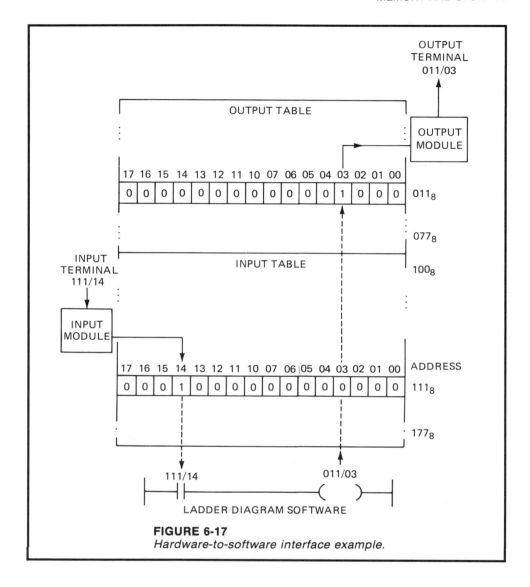

FIGURE 6-17
Hardware-to-software interface example.

numerous and random software errors. Proper documentation is also an aid in software troubleshooting and making software modifications.

A programmer will generally use three different documentation forms: (1) an overall data table map, (2) a data table word map, and (3) a data table bit assignment.

Overall Data Table Map   The data table map is used to list the addresses of groups of data words and to give a general description of the function of each group. The groups can include I/O image tables, block transfer, timer/counter, data files, and so on.

The forms are normally supplied by the PC vendor with pre-numbered rows representing word addresses. A typical example is shown in Figure 6-18. This form has pre-numbered rows listing addresses from 000 to 777. Each row lists 16 words where each space represents a one word address. Any group of related word addresses can be designated on the map by labeling the spaces representing their addresses.

Data Table Word Map   The data table word map is used to write functional descriptions of word addresses used for I/O image tables, word storage, timers and counters, and so on. A typical data word assignment sheet is shown in Figure 6-19.

In this example, the form is divided into two 64-word columns so the right-hand column can be numbered 100_8 greater than the left-hand column. This is useful because the standard counters and timers used in a PC system have the accumulated (ACC) value and the preset value stored in memory location separated by 100_8.

Data Table Bit Assignment   The data table bit assignment form is used to list the functions of input, output, and storage bits. It is normally similar to the word assignment sheet in that it is divided into two 2-word columns. The words are numbered consecutively for the convenient listing of the I/O bits having the same module group number.

For example, a portion of a typical data table bit assignment sheet is

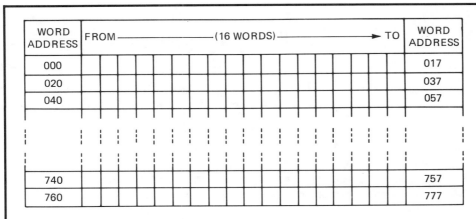

FIGURE 6-18
Overall data table word map.

WORD ADDRESS	DESCRIPTION	WORD ADDRESS	DESCRIPTION
030	PUMP CYCLE TIMER— ACC	130	PUMP CYCLE TIMER— PR
031	WASH TIME— ACC	131	WASH TIME— PR
032		132	
⋮	⋮		
076		176	
077		177	

FIGURE 6-19
Data word assignment.

WORD	BIT	DESCRIPTION	WORD	BIT	DESCRIPTION
013	00	WATER INLET VALVE	113	00	WATER INLET OPEN
013	01	WATER OUTLET VALVE	113	01	WATER OUTLET OPEN
013	02	PUMP 1 ON	113	02	PUMP RUNNING
⋮	⋮	⋮	⋮	⋮	⋮
013	16	HEATER 1 ON	113	16	TEMP. HI. TANK 4
013	17	HEATER 2 ON	113	17	TEMP. HI. PUMP A

FIGURE 6-20
Data table bit assignment.

shown in Figure 6-20. It illustrates listing the input and output devices associated with rack 1 and module group 3.

The bit assignment is a valuable aid in troubleshooting software and hardware problems throughout the design and start-up process.

EXERCISES

6.1 Describe the function and purpose of the four main parts of programmable controller memory.

6.2 List the advantages and disadvantages of the following memory types: RAM, EPROM, ROM, and core.

6.3 Determine memory size needed for a programmable controller system with 235 input points and 195 outputs (assume 30% spare memory capacity).

6.4 List the memory bit addresses for the following I/O points:
(a) an input to rack 1, module group 3, terminal 1
(b) an output from rack 2, module group 1, terminal 7
(c) an input to rack 2, module group 2, terminal 13

6.5 Draw an overall memory map for a PC system with 3 I/O racks, 10 counters, 15 timers, and 10 storage words (assume the process work area is from word 000 to 007).

6.6 Describe the functions of the system memory and the application memory in a PC system.

6.7 On the figure shown below circle the following I/O points:
(a) 01014, (b) 11401, (c) 01000, and (d) 11417.

17	16	15	14	13	12	11	10	07	06	05	04	03	02	01	00	
0	0	0	0	0	0	0	0	0	0	0	0	1	0	0	0	010_8
1	1	0	1	1	1	1	0	0	0	1	1	1	0	1	1	011_8

1	0	1	0	1	1	1	0	1	1	1	1	1	1	1	0	113_8
0	0	0	1	0	0	0	0	0	0	0	0	0	0	0	0	114_8

BIBLIOGRAPHY

1. Gilbert, R. A., and Llewellyn, J. A., *Programmable Controllers: Practices and Concepts,* Industrial Training Corporation (ITC), 1985.

2. Jones, C. T., and Bryan, L. A., *Programmable Controllers, Concepts and Applications,* International Programmable Controls, Inc., 1983.

3. Hughes, T. A., *Basics of Measurement and Control,* Instrument Society of America, 1988.

4. *Processor Manual PLC-5 Family Programmable Controllers,* Allen-Bradley Co., Inc., 1987.

5. *Programming and Operations Manual, PLC-2/30 Programmable Controllers,* Allen-Bradley Co., Inc., 1988.

6. *Understanding Computers: Memory and Storage,* Time-Life Books Inc., 1987.

7. Boylestad, R. L., and Nashelsky, L., *Electronic Devices and Circuit Theory,* Third Edition, Prentice-Hall, Inc. 1982.

7

Basic Programming Languages

Introduction   The programming language allows the user to communicate with the programmable controller (PC) via a programming device. PC manufacturers use several different programming languages, but they all convey to the system, by means of instructions, a basic control plan.

The four most common types of languages encountered in programmable controller system design are: (1) ladder diagrams, (2) Boolean mnemonics, (3) function blocks, and (4) the sequential function chart. These languages can be grouped into two major categories. The first two, ladder and Boolean, form basic PC languages, while function blocks and sequential function charting are considered high-level languages. The basic programmable controller languages consist of a set of instructions that will perform the most primitive type of control functions: relay replacement, timing, counting, sequencing, and logic. However, depending on the controller model, the instruction set may be extended or enhanced to perform other basic operations. The high-level languages have been brought about by a need to execute more powerful instructions that go beyond the simple timing, counting, and ON/OFF control. High-level languages are used for analog control, data manipulation, reporting, and other functions that are not possible with the basic instruction sets. These higher level languages will be discussed in the next chapter.

The language used in a PC actually dictates the range of applications in which the controller can be applied. Depending on the size and capabilities of the controller, one or more languages may be used. Typical combinations of languages are:

1. Ladder diagrams only
2. Boolean only
3. Ladder diagrams and function blocks

    4. Ladder and sequential function chart

    5. Ladder, function blocks, sequential function chart

Ladder Language
The ladder diagram language is a symbolic instruction set that is used to create a programmable controller program. It is composed of six categories of instructions that include relay-type, timer/counter, data manipulation, arithmetic, data transfer, and program control. The ladder instruction symbols can be formatted to obtain the desired control logic that is to be entered into memory.

The main function of the ladder diagram program is to control outputs based on input conditions. This control is accomplished through the use of what is referred to as a ladder rung. Figure 7-1 shows the basic structure of a ladder rung. In general, a rung consists of a set of input conditions represented by relay contact-type instructions and an output instruction at the end of the rung represented by the coil symbol. Throughout this section, the contact instructions for a rung may be referred to as input conditions, rung conditions, or control logic.

Coils and contacts are the basic symbols of the ladder diagram instruction set. The contact symbols programmed in a given rung represent conditions to be evaluated in order to determine the control of the output; all outputs are represented by coil symbols.

When programmed, each contact and coil is referenced with an address number that identifies what is being evaluated and what is being controlled. Recall that these address numbers reference the data table location of either an internal bit or a connected input or output. A contact, regardless of whether it represents an input/output connection or an internal bit, can be used throughout the program whenever that condition needs to be evaluated.

The format of the rung contacts is dependent on the desired control logic. Contacts may be placed in any configuration such as series, parallel, or series/parallel that is required to control a given output. For an output to be activated or energized, at least one left-to-right path of contacts must be closed. A complete closed path is referred to as having logic continuity. When logic conti-

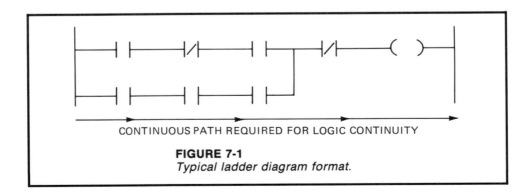

CONTINUOUS PATH REQUIRED FOR LOGIC CONTINUITY

FIGURE 7-1
Typical ladder diagram format.

nuity exists in at least one path, it is said that the rung condition is TRUE. The rung condition is FALSE if no path has continuity.

In the early years, the standard ladder instruction set was limited to performing only relay equivalent functions, using the basic relay-type contact and coil symbols similar to those illustrated in Figure 7-1. A need for greater flexibility, coupled with developments in technology, led to extended ladder diagram instructions that perform data manipulation, arithmetic, and program flow control. It is now commonplace to find controllers that include computer-like expressions similar to the diagram in Figure 7-2; the processor gets word 200 and puts it in memory word 201.

Relay Type Instructions   The relay-type instructions are the most basic of programmable controller instructions. They provide the same capabilities as hardwired relay logic discussed in Chapter 4, but with greater flexibility. These instructions primarily provide the ability to examine the ON/OFF status of a specific bit addressed in memory and to control the state of an internal or external output bit. The following is a description of relay-type instructions that are most commonly available in any controller that has a ladder diagram instruction set.

Normally Open Contact ⊣ ├─

The normally open contact is programmed when the presence of the input signal is needed to turn an output ON. When evaluated, the referenced address is examined for an ON (1) condition. The referenced address may represent the status of an external input, external output, or internal output. If, when examined, the referenced address is ON, then the normally open contact will close and allow logic continuity (power flow). If it is OFF (0), then the normally open contact will assume its normal programmed state (open), thus breaking logic continuity.

Normally Closed Contact ─┤ ╱ ├─

The normally closed contact is programmed when the absence of the referenced signal is needed to turn an output ON (1). When evaluated, the referenced address is examined for an OFF (0) condition. The referenced address may represent the status of an external input, external output, or an internal output. If, when examined, the referenced address is OFF, then the normally closed contact will remain closed, allowing logic continuity. If the referenced

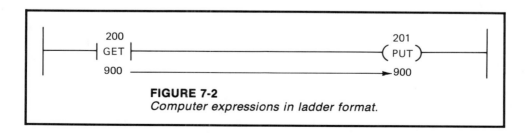

FIGURE 7-2
Computer expressions in ladder format.

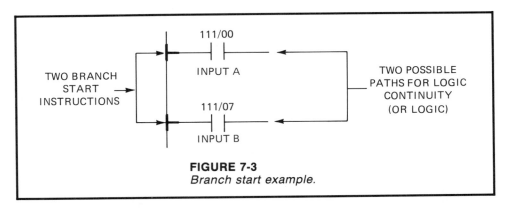

FIGURE 7-3
Branch start example.

address is ON, then the normally closed contact will open and break logic continuity.

Branch Start |—

The branch start instruction begins each parallel logic branch of a rung. It is the first instruction programmed if a parallel branch or logical OR function is needed in a logic rung. For example, if we need to perform the logical OR of input A (111/00) and input B (111/07) we would begin with a branch start as shown in Figure 7.3.

Branch End ⊤

The branch end instruction finishes a set of parallel branches. This instruction is used after the last instruction of the last branch to complete a set of parallel branches. To complete the continuity path for the OR circuit of Figure 7-3, we need to add a branch end instruction as shown in Figure 7-4.

Energize Coil —( )—

The energize coil instruction is programmed to control either an output connected to the controller, or an internal output bit. If any rung path has logic continuity, the referenced output is turned ON. The output is turned OFF if

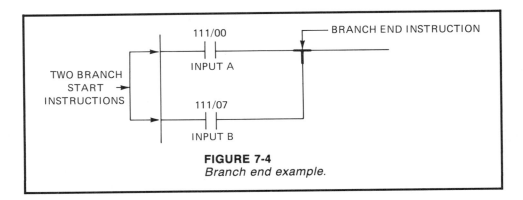

FIGURE 7-4
Branch end example.

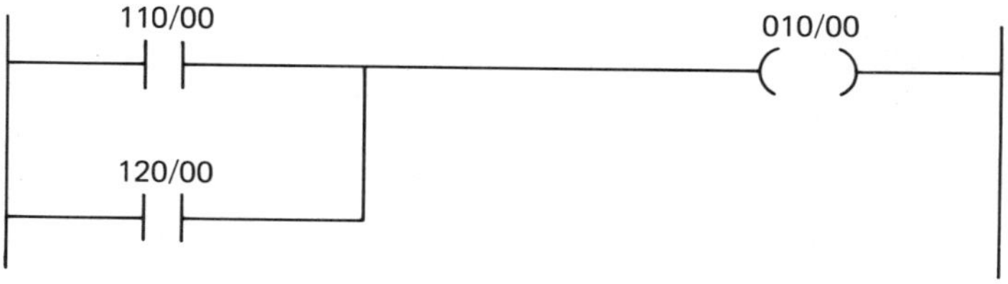

FIGURE 7-5
Example OR ladder logic program.

logic continuity is lost. When the output is ON, a normally opened contact of the same address will close, and a normally closed contact will open. If the output goes OFF, any normally opened set of contacts will then open, and normally closed contacts will close.

An example of a rung using relay-type instructions is shown in Figure 7-5. In this example, if either input 110/00 or input 120/00 is true, the output 010/00 will be true.

Several example problems will help to illustrate ladder logic programming.

EXAMPLE 7-1

Problem: Write a ladder diagram program to start and stop a pump. In this application, a local panel-mounted start push button is wired to input 110/00, and the stop push button is connected to input 110/01. The pump starter relay is connected to PC output 010/00 and the auxiliary (Aux.) motor start contacts (NO) are connected to PC input 110/02. Assume the START PB is normally open and the STOP PB is normally closed.

Solution: The application can be performed using the following program:

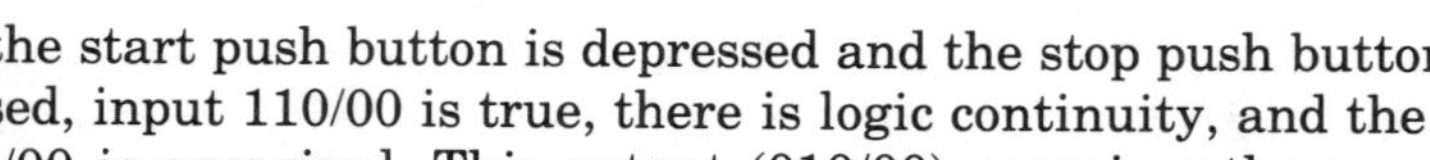

When the start push button is depressed and the stop push button is not depressed, input 110/00 is true, there is logic continuity, and the output bit 010/00 is energized. This output (010/00) energizes the pump start relay and the pump aux. switch is closed. This, in turn, sets input bit 110/02 ON to seal in the start push button (PB) and hold the pump ON until the stop PB is depressed.

EXAMPLE 7-2

Problem: There is only a single field-mounted push button available to start and stop an electric beacon. Write a ladder program to control the beacon. Assume the start/stop push button is wired to input point 110/06, and the starter relay for the beacon is connected to output 010/02.

Solution: The motor can be controlled using the following program.

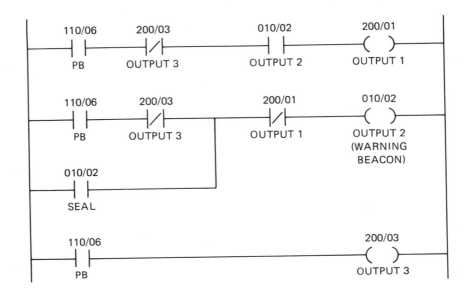

In this example program, when the push button (PB) is depressed for the first time, output coil 2 is energized and turns ON the warning beacon. The output 2 control bit (010/02) seals itself in. If the push button is depressed again, output 2 turns off. The second rung of the program detects the first time the push button is depressed, while the first rung senses the second time the push button is depressed.

Latch Coil ——(L)——

The latch coil instruction is programmed, if it is necessary, for an output to remain energized even though the status of the input bits that caused the output to energize may change. If any rung path has logic continuity, the output is turned ON and retained ON even if logic continuity or system power is lost. The latched output will remain latched ON until it is unlatched by an output instruction of the same reference address. The unlatch instruction is the only automatic (programmed) means of resetting the latched output. Although most controllers allow latching of internal or external outputs, some are restricted to latching internal outputs only.

Unlatch Coil —(u)—

The unlatch coil instruction is programmed to reset a latched output of the same reference address. If any rung path has logic continuity, the referenced address is turned OFF. The unlatch output is the only automatic means of resetting a latched output. Figure 7-6 illustrates the use of the latch and unlatch coils.

In this example, bit 100/00 must be true to set latch bit 200/00. If this bit (100/00) goes false, the output coil will remain ON until bit 100/01 is true to unlatch the coil. Note that the output unlatch instruction addresses the same memory bit that has been latched ON.

Timer and Counter Instructions

Timers and counters are output instructions that provide the same functions as would hardware timers and counters. They are used to activate or deactivate a device after an expired interval or count. The timer and counter instructions are generally considered internal outputs. Like the relay-type instructions, timer and counter instructions are fundamental to the ladder diagram instruction set.

The operations of timers and counters are quite similar, in that they are both counters. A timer counts the number of times that a fixed interval of time (e.g., 0.1 sec, 1.0 sec) elapses. To time an interval of 3 seconds, a timer counts three 1-second intervals (time base). A counter simply counts the occurrence of an event. The timer and counter instructions require an accumulator (ACC) register (word location) to store the elapsed count and a preset (PR) register to store a preset value. The preset value will determine the number of event occurrences or time-based intervals that are to be counted. When the accumulated value equals the preset value, a status bit is set on and can be used to turn on an output device.

Since timer or counter instructions store an accumulated and preset value, two words of a data table are required for these instructions. Normally, the accumulated and preset values are stored in memory in 3-digit binary coded decimal (BCD) format. The BCD numbers can range from 000 to 999 and are stored in the lower 12 bits of a data word (see Figure 7-7). Each BCD digit is represented by a 4-bit group of binary numbers. This arrangement of 1s and

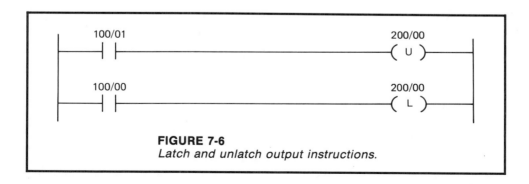

FIGURE 7-6
Latch and unlatch output instructions.

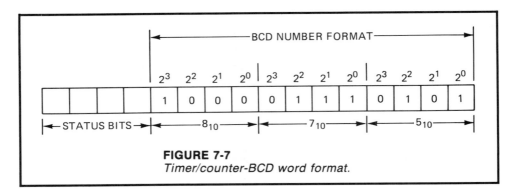

FIGURE 7-7
Timer/counter-BCD word format.

0s in a group of four bits corresponds to a decimal number from 0 to 9, as discussed in Chapter 2.

The leftmost four bits in the timer/counter ACC and PR word (bits 14–17) are not used to form a BCD number. In the accumulated word, they are used as status bits. Bit 15 is either set ON or OFF, depending on the type of timer or counter instruction, when timing or counting is complete. For all timers/counters in this presentation, bit 17 is set ON when rung conditions are true and is set OFF when the rung logic is false.

A typical timer accumulated word is shown in Figure 7-8. The timer DONE bit is bit 15 and the timer ON bit is bit 17. The timer done contact can be used in the program as a normally open (NO) or normally closed (NC) contact to perform some control or logic operation.

Time Delay Energized (ON) —(TON)—

The time delay ON timer output instruction is programmed to provide time delayed action or to measure the duration for which some event is occurring. If any rung path has logic continuity as shown in Figure 7-9, the timer begins

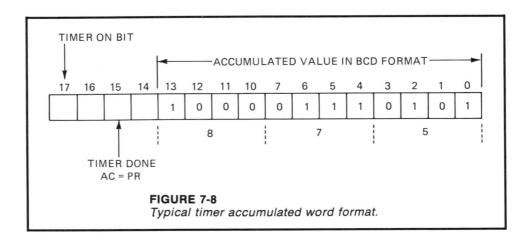

FIGURE 7-8
Typical timer accumulated word format.

counting time-based intervals and counts until the accumulated (ACC) time equals the preset (PR) value as long as the rung conditions remain true. When the accumulated time equals the preset time, a timer DONE bit in the accumulated word is set to 1. Whenever the rung logic conditions for the TON instruction go false, the accumulated value is reset to all zeros.

In the example application shown in Figure 7-9, when the pump start switch is ON, the timer begins to count time-base intervals. As long as the switch remains closed or ON, the timer increments its accumulated value word

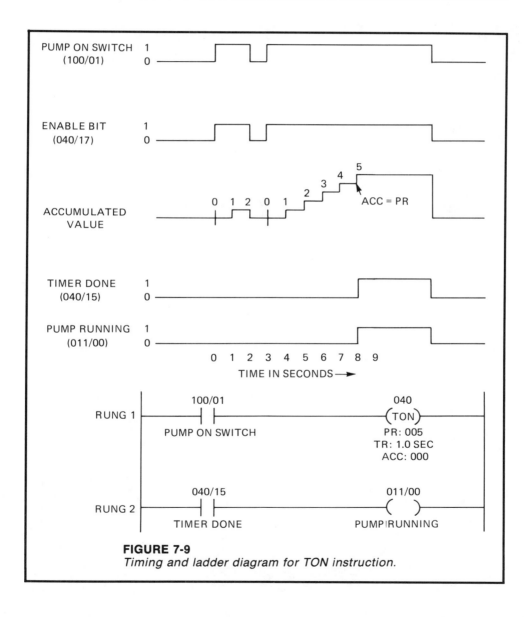

FIGURE 7-9
Timing and ladder diagram for TON instruction.

for each counted interval. When the accumulated value equals the programmed preset value of 5 seconds, the timer stops incrementing its accumulated value and sets the timer DONE bit of word 040. This DONE bit is then used in rung 2 to energize the pump running output bit (011/00).

Timer Off Delay —(TOF)—

The timer off delay output instruction is programmed to provide time delayed action. If logic continuity is lost, the timer begins counting time-based intervals

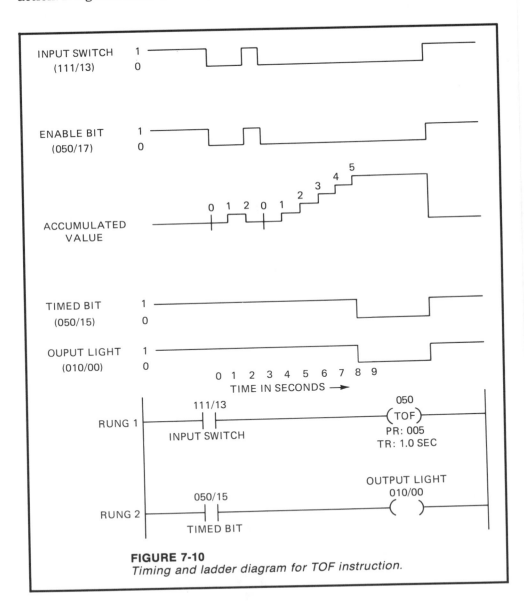

FIGURE 7-10
Timing and ladder diagram for TOF instruction.

until the accumulated time equals the programmed preset value. When the accumulated time equals the preset time, the output is de-energized, and the timed bit (bit 15) is set to zero. The timed contact can be used throughout the program as a NO or NC contact. If logic continuity is gained before the timer is timed out, the accumulator word is set to zero. A timing diagram and example program for a TOF instruction with a preset value of 5 seconds are shown in Figure 7-10.

In rung 1, if the input switch is open (logic 0) the timer will count up to

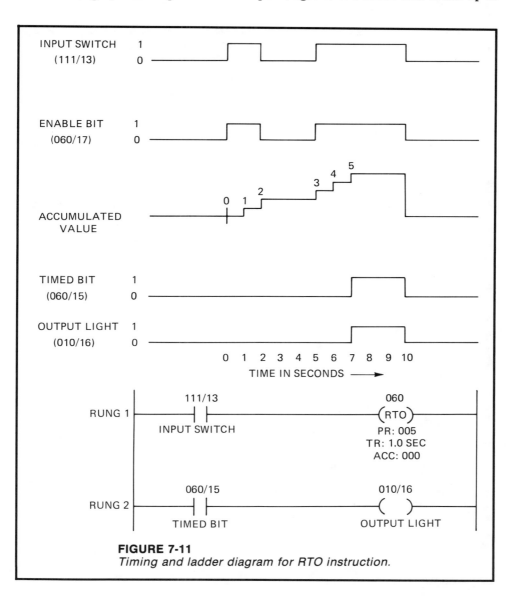

FIGURE 7-11
Timing and ladder diagram for RTO instruction.

5. When the preset value is equal to the accumulated value, the timed bit (050/15) will be set to 0 and the output light will turn off in the second rung.

Retentive ON-Delay Timer —(RTO)—

The retentive timer output instruction is used if it is necessary for the timer accumulated value to be retained, even if logic continuity or power is lost. If the timer rung path has logic continuity, the timer begins counting time-based intervals until the accumulated time equals the preset value. The accumulator register retains the accumulated value, even if logic continuity is lost before the timer is timed out or if power is lost as shown in Figure 7-11. When the accumulated time equals the preset time, the output is energized, and the timed out contact associated with the output is turned ON. The timer contacts can be used throughout the program as a NO or NC contact. The retentive timer accumulator value must be reset by the retentive timer reset instruction.

Retentive Timer Reset —(RTR)—

The retentive timer reset output instruction is the only automatic means of resetting the accumulated value of a retentive timer. If any rung path has logic continuity, then the accumulated value of the retentive timer with the same word address is reset to zero.

EXAMPLE 7-3

Problem: Write a ladder program and draw the timing diagram using a retentive timer at word location 060 with an input bit at 111/13 that turns on a solenoid valve at an output address of 010/00 after a 1/2-second time delay. Use input switch 111/01 to reset the timer.

Solution:

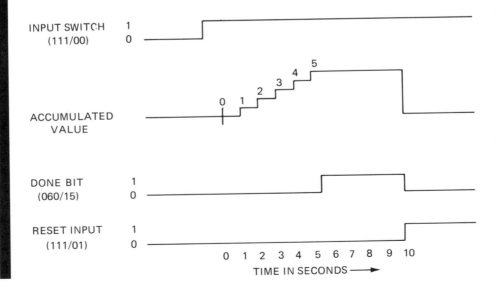

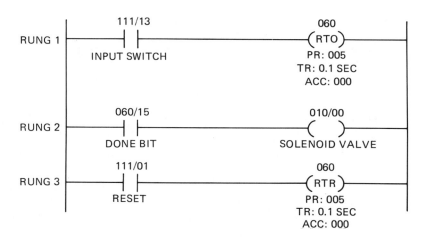

EXAMPLE 7-4

Problem: Design a simple timing circuit that can be used to generate a square wave pulse of any width.

Solution: The ladder logic program consists of two TON instructions as shown.

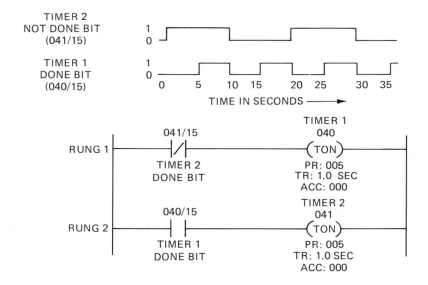

Up-Counter ——(CTU)——

The up-counter output instruction will increment by one each time the counted event occurs. A control application of a counter is to turn a device ON or OFF after reaching a certain count. An accounting application of a counter is to keep track of the number of filled cans that pass a certain point. The up-counter

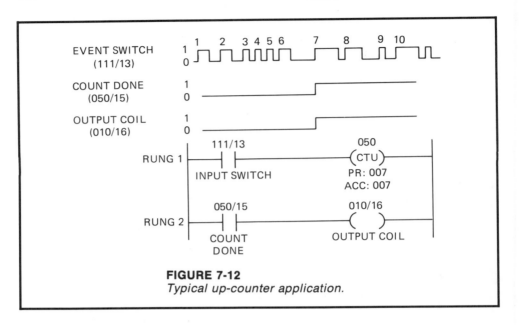

FIGURE 7-12
Typical up-counter application.

increments its accumulated value each time the up-count logic input makes an OFF-to-ON transition. Since only the false-to-true transition causes a count to be accepted, the rung condition must go from true to false and back to true before the next count is registered. Figure 7-12 shows a typical up-counter application.

When the accumulated value reaches the preset (PR) value, the output is turned ON, and the COUNT DONE bit (050/15) is used to activate an output coil (010/16).

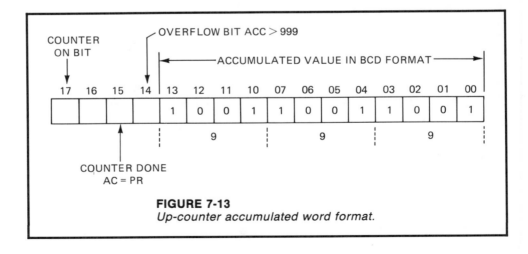

FIGURE 7-13
Up-counter accumulated word format.

Unlike a timer instruction, the counter instruction continues to increment its accumulated value after the preset value has been reached. If the accumulated value goes above the maximum range of the counter, normally 999, an overflow bit will be set. Status bit 14 is used as the overflow bit as shown in Figure 7-13. This overflow bit can be used to cascade counters for counter applications greater than 999.

Counter Reset —(CTR)—

The counter reset output instruction is used to reset the up- and down-counter accumulated values. When programmed, the CTR coil is given the same reference address as the CTU and CTD coils. The preset and accumulated values are displayed on the ladder diagram, but they have no real function. If the CTR rung condition is TRUE, the counter with the same address will be cleared.

EXAMPLE 7-5

Problem: Write a ladder logic program to count up to 25 events and then reset the counter to zero.

Solution:

```
                       111/13                                    050
RUNG 1  ├──────────────┤ ├──────────────────────────────────(CTU)──────┤
                       EVENT                                  PR: 025
                       SWITCH                                 ACC: 000

                       050/15                                    050
RUNG 2  ├──────────────┤ ├──────────────────────────────────(CTR)──────┤
                       DONE BIT                               PR: 025
                                                             ACC: 000
```

Down-Counter —(CTD)—

The down-counter output instruction will count down by one each time a certain event occurs. Each time the down-count event occurs, the accumulated (ACC) value is decremented. In normal use, the down-counter is used in conjunction with the up-counter to form an up/down counter.

EXAMPLE 7-6

Problem: Design an up/down counter circuit to count the number of parts produced on an assembly line. Assume input 111/13 is activated by each part produced in final assembly, input 110/01 is activated by a part being rejected in final test, and input 110/02 is energized at the end of a production run.

Solution:

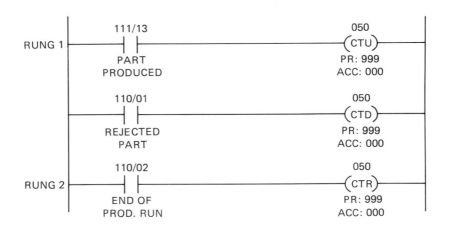

Data Transfer Operations   Data transfer instructions involve the transfer of the contents from one register to another. Data transfer instructions can address any location in the memory data table, with the exception of areas restricted to user application. Prestored values can be automatically retrieved and placed in any new location. That location may be the preset register for a timer or counter or even an output register that controls a seven-segment display.

Get Word —(GET)—
The GET instruction accesses the contents of the referenced address and makes it available for other operations. Some controllers use this instruction to access registers to perform math operations or comparisons.

Put Word —(PUT)—
The PUT instruction is used to store the result of other operations in the memory location (register) specified by the PUT coil. PUT is generally used with the GET instruction to perform a move operation.

Arithmetic Operations   The arithmetic operations include the four basic operations of addition, subtraction, multiplication, and division. These instructions use the contents of two word locations and perform the desired function.

The arithmetic instructions are programmed in the output portion of the ladder and use either one or two data words to store the result. The add and subtract instructions use one word. Multiply and divide need two words for the computed result.

The result is normally stored in BCD format in the lower 12 bits of the arithmetic instruction word as shown in Figure 7-14.

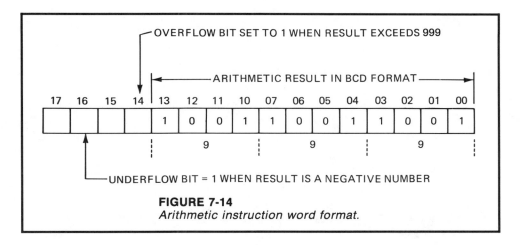

FIGURE 7-14
Arithmetic instruction word format.

Addition (ADD) —(+)—

The ADD instruction performs the addition of two values stored in two different memory locations. The processor uses a GET (data transfer) instruction to access the two values. The result is stored in the word address referenced by the ADD coil as shown in Figure 7-15. If the addition operation is enabled only when the rung conditions are true, then the input conditions should be programmed before the values are accessed in the addition rung. Overflow conditions are usually signaled by one of the bits of the addition result register.

Subtraction (SUB —(–)—

The SUB instruction performs the subtraction operation of two registers. As in addition, if there is a condition to enable the subtraction, it should be programmed before the values are accessed in the rung. The subtraction result register will use a minus sign to represent a negative result as shown in Figure 7-16.

Multiplication (MUL) —(×)—(×)—

The MUL instruction performs the multiplication operation. It uses two registers to hold the results of the operation between two operand registers. The

FIGURE 7-15
Typical addition logic rung.

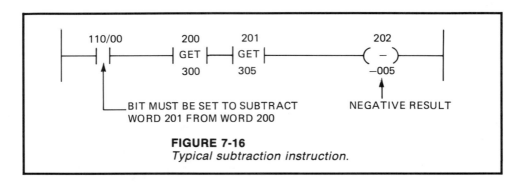

FIGURE 7-16
Typical subtraction instruction.

two registers are referenced by two output coils. If there is a condition to enable the operation, it should be programmed before the two operands are accessed in the multiplication rung.

Normally, the systems programmer will choose two consecutive memory words to store the product as shown in Figure 7-17. If the result is less than 6 digits, leading zeros will normally be displayed in the product.

Divide (DIV) ─(÷)─(÷)─

The DIV instruction performs the quotient calculations of two numbers. The result of the division is held in two result registers as referenced by the output

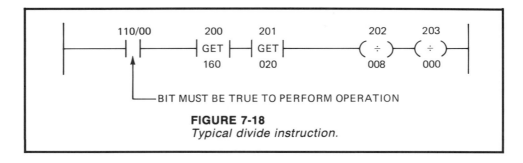

FIGURE 7-17
Typical multiply instruction.

FIGURE 7-18
Typical divide instruction.

coils. The first result register generally holds the integer, while the second result register holds the decimal fraction as shown in Figure 7-18.

A typical math application in programmable controller systems is to scale process analog input signals, as illustrated in the next example.

EXAMPLE 7-7

Problem: An analog input signal from a pressure transmitter has a range of 0 to 20 inches. This signal is converted to 000 to 999 in the analog input module. Write a ladder logic program to convert the signal back to engineering units and display using BCD output module.

Solution: The first step is to calculate the scale factor as follows:

$$\text{Scale Factor} = \frac{\text{Pressure Range}}{\text{PC Range}}$$

$$= \frac{(20 - 0) \text{ in.}}{999 - 0}$$

$$= 0.02 \text{ in.}$$

We now use this scale factor in the ladder diagram as shown. Since decimal points are normally not allowed in programmable controller words, the scale factor is changed to 20 so that the pressure data is stored in the left result word (202) to obtain the correct answer.

Data Comparison Operations   In general, the manipulation of data using ladder diagram instructions involves simple register (word) operations to compare the contents of two registers. In the ladder language, there are three basic data comparison instructions: equal to, greater than, and less than. Based on the result of a greater than, less than, or equal to comparison, an output can be turned ON or OFF, or some other operation can be performed.

Equal to ─┤ = ├─

The equal to instruction is used to compare the contents of two referenced registers for an equal condition, when the rung conditions are true. As in the arithmetic instructions, the values to be compared can be accessed directly or through a GET instruction depending on the controller. If the operation is

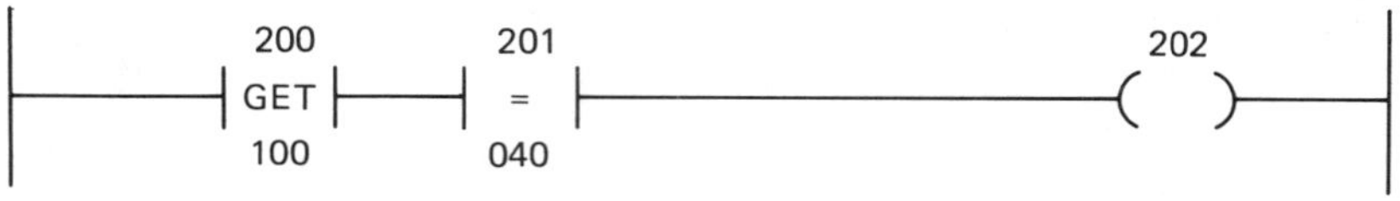

FIGURE 7-19
Typical "equal to" instruction.

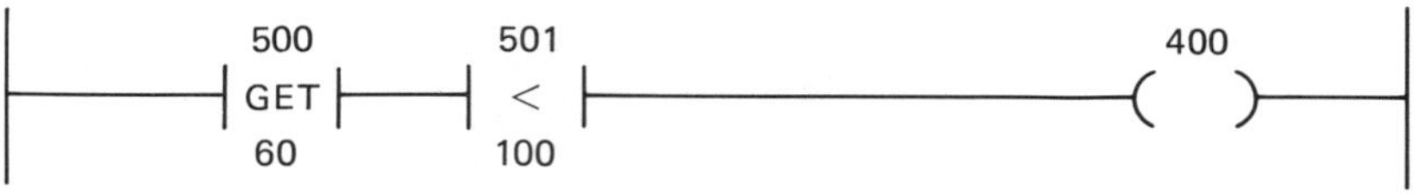

FIGURE 7-20
Typical "less than" instruction.

true, the output coil is energized. A typical "equal to" programming example is shown in Figure 7-19. In this example, when the GET value in word 200 equals the value in the "equal" word 201, the output coil is energized.

Less Than ─| < |─
Similar to the equal to instruction, the less than instruction tests the contents of the value of one register to see if it is less than the value stored in a second register. If the test condition is true, the output coil is energized as shown in Figure 7-20.

Greater Than ─| > |─
The greater than instruction operates like the less than operation, with the exception that the test is performed for a greater than condition. If the test condition is true, the output coil is energized. Some controllers do not have this function, since you can perform a "greater than" function using the "less than" logic by reversing the order of the data and the "less than" function in the logic rung.

EXAMPLE 7-8

Problem: Write a ladder logic program to energize output coil 010/00, if the data in word 400 is less than or equal to the data in word 401 when input 111/15 is true.

Solution:

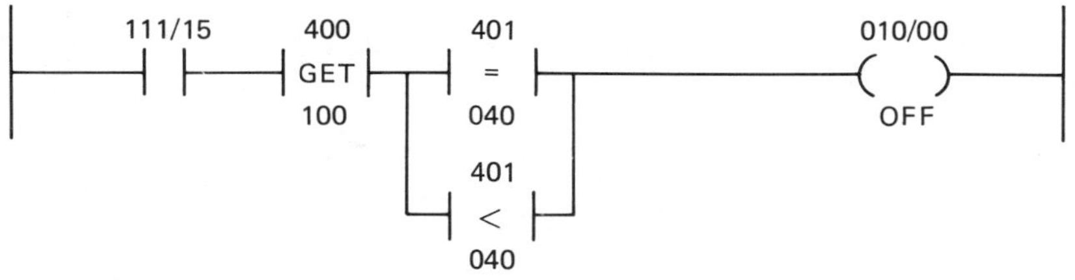

Figure 7-21 shows an example ladder diagram using arithmetic, data manipulation, and data transfer instructions. In this figure, if input 111/04 is true, the data in memory location 1600 is compared to the data value in location 1601; if they are equal, output coil 100/07 is energized. When output bit 100/07 is energized, the data in memory locations 1700 and 1701 are added and stored in location 1702. Finally, if the logic bit 111/16 is ON, the contents of 1702 are transferred to output word 500.

Program Control Instructions   The program control functions are used to perform a series of conditional and unconditional jump and return instructions. These instructions allow the program to execute only certain sections of the control logic if a fixed set of logic conditions are met. The following instructions are a representative selection of some of the program control instructions available in most controllers.

Master Control Relay —(MCR)—
The MCR output instruction is used to activate or deactivate the execution of a group or zone of ladder rungs. The MCR is used in conjunction with an END

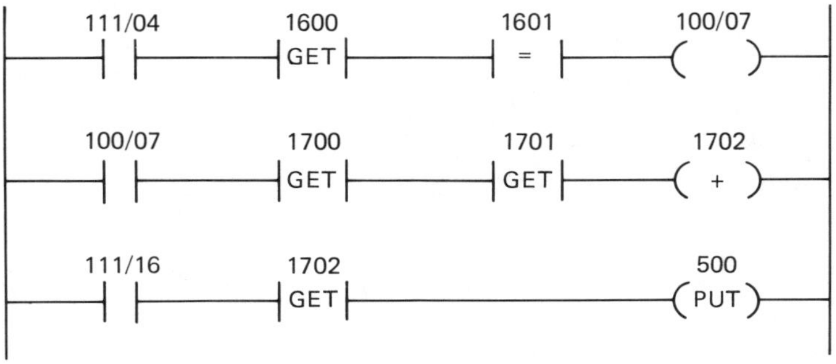

FIGURE 7-21
Example arithmetic and data manipulation program.

MCR coil to place a fence around the group of rungs. For example, in Figure 7-22, if the STOP input is active, the MCR coil will be energized and the logic inside the zone will be executed by the controller. If the MCR coil is turned OFF, all outputs inside the zone will be deenergized.

Jump —(JMP)—

The jump instruction allows the normal sequential execution to be altered so that the CPU will jump to a new position in the ladder program. If the jump rung logic is true, the jump coil (JMP) instructs the CPU to jump to and execute the rung labeled with the same reference address as the jump coil. This allows the program to execute rungs out of the normal sequential flow of a standard ladder program.

Label —[LBL]—

The label (LBL) is to identify that ladder rung which is the destination of a jump instruction. The label reference must match that of the jump instruction with which it is used. The label instruction does not contribute to logic continuity, and it is always logically true. It is placed as the first logic condition in the rung. A label instruction referenced by a unique address can be defined only once in a program.

Return —(RET)—

The return (RET) instruction is used to terminate a ladder jump subroutine. The output coil is programmed normally without any rung logic inputs. When

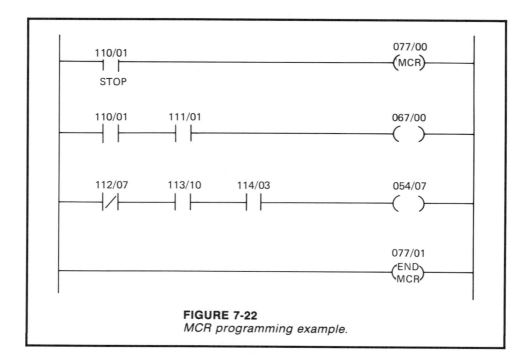

FIGURE 7-22
MCR programming example.

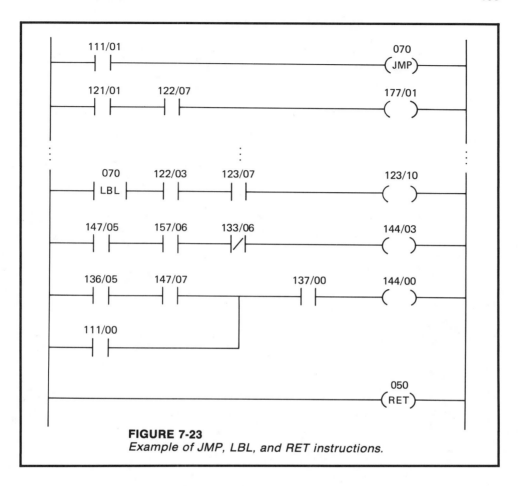

FIGURE 7-23
Example of JMP, LBL, and RET instructions.

a RET coil is encountered, the program returns to the spot where it left the main program and begins at the ladder rung immediately following the jump instruction that initiated the subroutine. Normal program execution continues from that point. A return must be used at the end of each subroutine.

The example program in Figure 7-23 illustrates the jump, label, and return program control instructions. If the logic input 111/01 is true, the program execution jumps to the subroutine label 070. After the subroutine is completed, program execution returns to the rung following the jump instruction.

Boolean Language
The Boolean language is a basic level PC language that is primarily based on Boolean operators such as AND, OR, and NOT. However, like ladder programming, it also uses mnemonic instructions, such as ADD (addition), TON (timer on), JMP (jump) and so on. A typical Boolean instruction set and their ladder diagram equivalents are given in Table 7-1. It is interesting to note that many of the mnemonic instructions

TABLE 7-1
Typical Boolean Set and Ladder Diagram Equivalents

Mnemonic	Function	Ladder Equivalent
LD	Load input	┤ ├
LD NOT	Load NC input	┤/├
AND	Logical AND	——
OR	Logical OR	⊐
OUT	Energize output	—( )—
TIM	Timer	—(TON)—
CNT	Counter	—(CTN)—
ADD	Addition	—(+)—
SUB	Subtraction	—(−)—
MUL	Multiplication	—(×)—

are the same as ladder language. The only real difference is the program entry format and listing.

The fundamentals of Boolean logic were discussed in Chapter 3. A example application will best illustrate Boolean programming.

EXAMPLE 7-9

Problem: Write a ladder program and its Boolean equivalent for the following control application: Turn on pump 1 (I/O address 010/00) if tank level is not high (111/00) and tank level is low (111/07).

Solution:

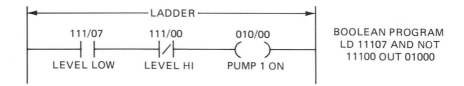

The Boolean language is generally only used in small systems with less than 100 input/output points. It is normally programmed using hand-held portable programmable controllers and the documentation is handwritten or typed, since the programs are relatively simple.

EXERCISES

7.1 Write a ladder diagram program to manually control an electric motor. Assume that a start push button is wired to an input module at 100/00 and a stop push button is connected to the same input module at address

100/01. Also assume that the start PB is normally open and the stop push button is normally closed. An output module point will drive a motor control relay at address 040/01, and there is an auxiliary motor start contact (NO) connected to PC input 100/02.

7.2 Revise the motor control ladder program shown using latch and unlatch output instructions instead of the standard output coil shown.

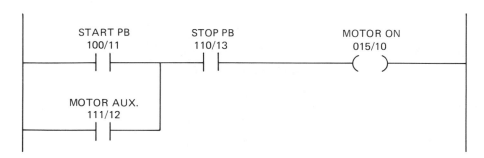

7.3 Write a ladder logic program to control a process pump under the following conditions:
   1. Turn on the pump (output 010/00) five seconds after the inlet valve (input 110/10) and the outlet valve to the pump have been opened.
   2. Turn off the pump if either the inlet or the outlet valve to the pump is closed.

7.4 Design a ladder program using timer instructions to flash four process alarm lights on a control panel at a rate of 0.3 second. The alarm lights are driven by PC outputs at 010/00, 010/01, 010/02, and 010/03. Connect a panel-mounted push button to input point 110/00 to acknowledge the alarms. Assume the alarm lights will remain ON after the acknowledge PB is depressed and will turn OFF only if the alarm input has been cleared. Do a control panel layout and I/O wiring diagrams for the system.

7.5 A temperature transmitter connected to an analog input module has a range of 100°C to 500°C. This signal is converted to 000 to 999 in BCD format in the programmable controller. Design a ladder logic program to convert the signal back to engineering units and display using a BCD output module. Draw a wiring diagram for the system showing the input and output wiring.

7.6 Design a ladder program to control three heater elements in a process fluid storage tank used to heat the fluid to 400°C. Assume that the heater contactors are energized at 1-second intervals to produce a smoother heat curve and better energy conversion. Use ON/OFF control to maintain the temperature at the set point.

7.7 Write a ladder logic program to transfer some BCD data in word 600 to BCD output display at word 700, if the data in word 600 is greater than the data in word 400 and less than the data in location 401.

7.8 Design a Boolean language program to manually control an electric pump. Assume the following: a normally open start push button is wired to an input module at 100/10, a normally closed stop push button is connected to input 100/11, a discrete output module drives a pump starter relay at address 040/02, and there is an auxiliary motor start contact (NO) connected to input point 100/12.

BIBLIOGRAPHY

1. Jones, C. T., and Bryan, L. A., *Programmable Controllers, Concepts and Applications,* International Programmable Controls, Inc., First Edition, 1983.

2. *Programmable Controller Fundamentals,* Allen-Bradley Company, 1985.

3. *PLC-2/30 Programmable Controller: Programming and Operations Manual,* Allen-Bradley, Publication 1772-6.8.3, 1984.

4. Hughes, T. A., *Basics of Measurement and Control,* Instrument Society of America, 1987.

5. Gilbert, R. A., and Llewellyn, J. A., *Programmable Controllers—Practices and Concepts,* Industrial Training Corporation, 1985.

8

High-Level Programming Languages

Introduction   The higher-level programmable controller languages are required to perform more powerful instructions beyond simple ON/OFF control, timing, counting, and data manipulation. These high-level languages are used for analog control, data file operations, data block transfer, sequencer operations, data reporting, and other functions that are not possible with the basic programming languages. The common high-level languages to be discussed are function block instructions and sequential flowcharting. Both languages are normally used in combination with ladder logic programming, so numerous examples will be included that combine ladder and high-level programming.

Function Block Instructions   Function blocks instructions allow the user to program more complex functions such as (1) data transfer file instructions, (2) block transfer read/write instructions, (3) shift register instructions, and (4) sequencer instructions.

Data Transfer File Instructions   A file is a group of consecutive data table words used to store PC information. A file is normally defined by a counter and the starting word address of the file. The counter performs two functions: It defines the number of words in the file (file length) with its preset value, and it points to a particular word in the file (position) with its accumulated value. The counter address is sometimes called the instruction address. It is the address used by the processor to search for the instruction. In most PCs, the words in the file must be located one after the other, and the files are normally restricted to 999 words in length.

   The structure of a typical file is shown in Figure 8-1. This figure illustrates a 7-word file starting at word address 200_8. The counter (at address

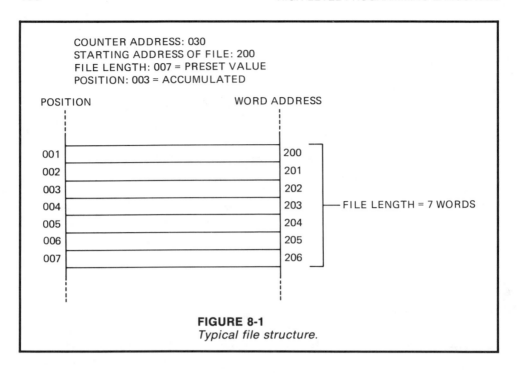

COUNTER ADDRESS: 030
STARTING ADDRESS OF FILE: 200
FILE LENGTH: 007 = PRESET VALUE
POSITION: 003 = ACCUMULATED

FIGURE 8-1
Typical file structure.

030_8) has an accumulated value of 003. So it is pointing to the third word in the file, word address 202_8. This word might be either the source or destination of the data that is currently being operated upon by the file instruction.

File instructions are displayed in block format on the programming terminal as in Figure 8-2.

The *counter address* is the address in memory of the accumulated value

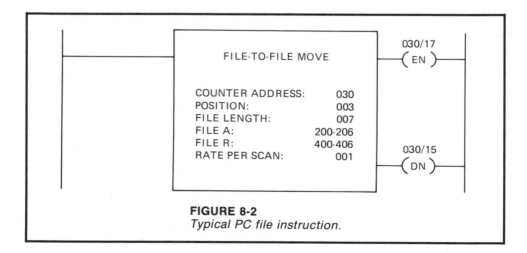

FILE-TO-FILE MOVE

COUNTER ADDRESS:	030
POSITION:	003
FILE LENGTH:	007
FILE A:	200-206
FILE R:	400-406
RATE PER SCAN:	001

030/17 (EN)

030/15 (DN)

FIGURE 8-2
Typical PC file instruction.

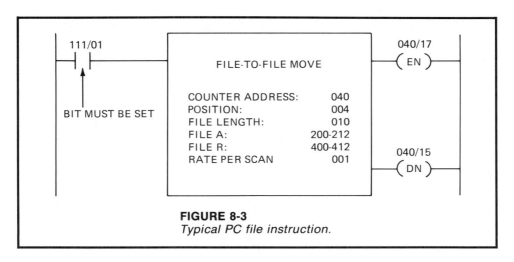

FIGURE 8-3
Typical PC file instruction.

of the instruction. It is also the address used by the processor to search for the instruction. The *position* is the current word being operated upon (accumulated value of counter). The *file length* is the number of words in the file (preset value of the counter). *File A* is the starting address of the source file. *File R* is the starting address of the destination file. The rate *per scan* is the number of data words moved per scan.

The output coils to the right of the file instruction are the enable (EN) and done (DN) bits. These bits have the same word address as the instruction counter. The address of these status bits is automatically set by the processor when the counter address is entered by the programmer. The enable bit is set high when the rung condition is true and/or when the instruction is being performed. The done bit is set high when the operation is complete and remains high as long as the rung condition is true.

The most common data transfer file instructions are (1) file-to-file move, (2) file-to-word move, and (3) word-to-file move.

File-to-File Move A file-to-file move is used to transfer a large amount of data at one time from one block location of memory (file A) to another location (file R). The files can normally be from 1 to 999 words long. File A remains unaffected by the operation. An example file-to-file move instruction is shown in Figure 8-3.

A typical programming example will illustrate the file-to-file move instruction.

EXAMPLE 8-1

Problem: Write a program to transfer 8 words of process data each second from memory locations 300 through 307 to another block of memory starting at word 400.

Solution: To perform the required operation, we need a 1-second timer and a file-to-file move instruction as shown.

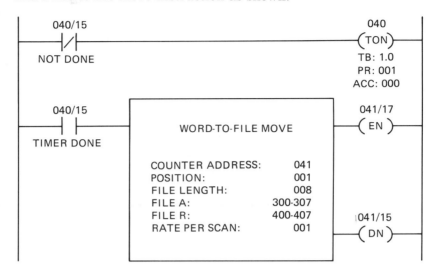

File-to-Word Move   The file-to-word move instruction transfers a duplicate of a word from a word location in file A to a specified word W in the data table. A typical example of a file-to-word move instruction is shown in Figure 8-4. In this example, words from memory locations 200 through 212 (file A) are moved one word at a time to word location 400 (word W) depending on the position of a pointer in the instruction. In this case, since the pointer is at 6, the sixth word, 205, will be transferred to word 400 when the instruction is executed.

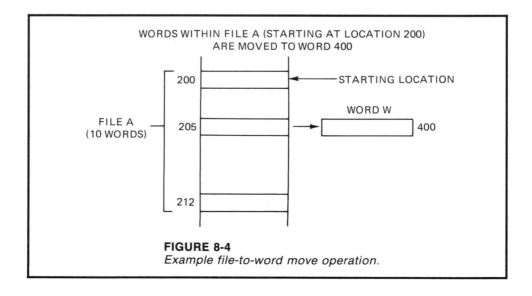

FIGURE 8-4
Example file-to-word move operation.

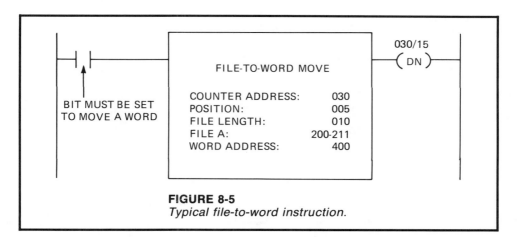

FIGURE 8-5
Typical file-to-word instruction.

To program the file-to-word operation, we use the instruction format shown in Figure 8-5.

The counter *address* is 030 for the typical example in Figure 8-5; this is the instruction address. The *position* or the current word being operated upon is 5, so the word that will be moved when the input logic is true will be word 204. That word, 204, will be transferred to word 400. The *file length* or the number of words in the file is 10 in this example. *File A* is located between words 200 and 211 octal or 10 words decimal, and the word *address* is 400. Data will be transferred to word 400 each time the input logic goes from false to true.

EXAMPLE 8-2

Problem: Write a program to transfer temperature data from 8 consecutive memory locations starting at word 200 to a panel-mounted LED display. Assume the operator uses a control panel push button at address 100/10 to request a temperature display. The LED is connected to the digital output module and the module uses memory word 400.

Solution: The panel-mounted push button is used by the operator to step the file-to-word instruction as shown.

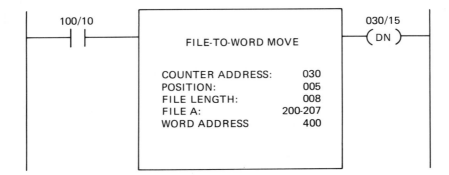

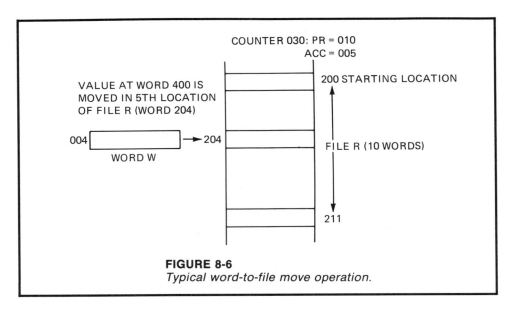

FIGURE 8-6
Typical word-to-file move operation.

Word-to-File Move The word-to-file move instruction transfers a duplicate of the value in a specified data table word, W, into a word in file R that is pointed to by the counter accumulated value. In the example of Figure 8-6, the data in word 400 is transferred in to a file word when input rung logic is set. In Figure 8-6 the position value is 5, so word 400 will be transferred to the fifth location of file R or word 204 when the instruction is executed by the processor.

The instruction format needed to perform the word-to-file operation is

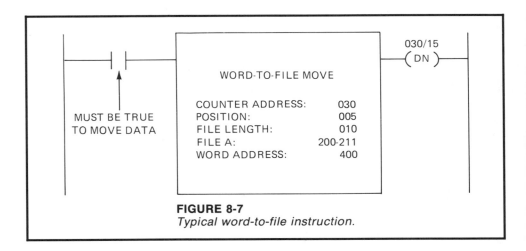

FIGURE 8-7
Typical word-to-file instruction.

given in Figure 8-7. In the figure, the counter address (030) is the memory address of the instruction in the programmable controller. The position is 5, so word 204 is being addressed. The file length is 10, so there are 10 words in the file. The word address is 400; data is transferred to this location on a false-to-true transition of the input logic. File R is the starting address of the destination file or 200 to 211. It is important to note that the word addresses are in octal, so the 10 words have addresses 200, 201, 202, 203, 204, 205, 206, 207, 210, and 211.

EXAMPLE 8-3

Problem: Write a program to transfer numbers from a 3-digit BCD thumbwheel switch to 10 consecutive memory locations in a programmable controller. Assume the operator uses a normally open (NO) push button at input location 100/10 to enter each BCD number; and the thumbwheel is connected to a BCD input module; and the module transfers the input data to memory word 400.

Solution: The push button is used by the operator to step the word-to-file instruction as shown. The file instruction then transfers the data to memory starting at word 200.

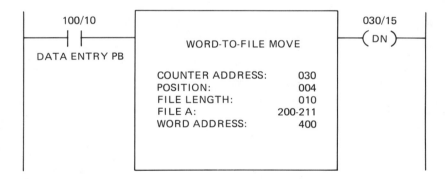

Block Transfer Read/Write Instructions The block transfer instruction is normally used to transfer up to sixty-four 16-bit words of data in one processor scan from I/O modules to or from the data table. It is used with intelligent I/O modules, such as PID, analog, ASCII, thermocouple, or encoder/counter modules. In most PCs, the block transfer instruction moves only one word data per scan.

The block transfer instruction can be performed as a read, write, or bi-directional operation, depending on the I/O module being used. An input module uses block transfer read, an output module uses a block transfer write, and a bi-directional module uses both the read and write operations. During a read operation, data is read into memory of the programmable controller system from the intelligent module. During a write operation, information is written to the output module from memory in the programmable controller system.

The programmable controller uses two I/O image table bytes (8 bits) to communicate with block transfer modules. The byte corresponding to the I/O module's address in the output image table (control byte) holds the read or write bit for initiating the transfer of data. The byte corresponding to the I/O module's address in the input image table (status byte) is used to indicate the completion of the data transfer.

Whether the upper or lower byte of the I/O image table word is used depends on the position of the module in the module group (i.e., slot 0 or slot 1). When in slot 0 (lower slot), the lower byte is used, and, when in slot 1 (upper slot), the upper byte is used (see Figure 8-8).

A typical format for a block transfer read instruction is shown in Figure 8-9. The module is located in rack 1, module group 0, and slot 1; therefore, the

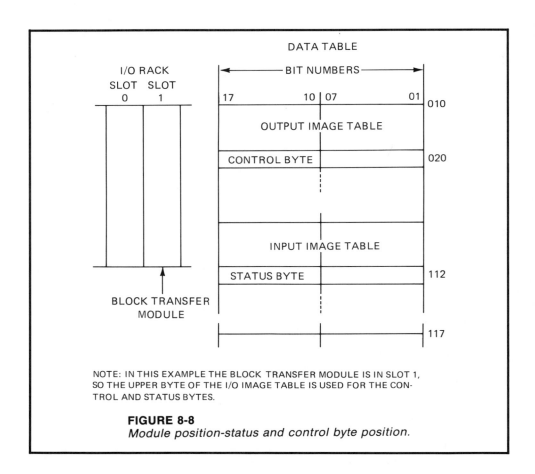

NOTE: IN THIS EXAMPLE THE BLOCK TRANSFER MODULE IS IN SLOT 1, SO THE UPPER BYTE OF THE I/O IMAGE TABLE IS USED FOR THE CONTROL AND STATUS BYTES.

FIGURE 8-8
Module position-status and control byte position.

control and status bytes corresponding to the address of the module are at word locations 010 and 110 (upper bytes), respectively. The address of the read enable bit is 010/17, and the address of the done bit is 110/17 for the read instruction as shown in Figure 8-9.

During the program scan when the rung logic is true, the read instruction is enabled and the enable bit is set to 1. In the next scan of the processor, the control byte at word 010 is sent to the intelligent module. This control byte contains the instructions to the module requesting data be sent to the processor. The module responds that it is ready to send data. The processor then searches the data table and finds module address 101 in word address 030 and the file address, 070, in word 130.

The processor then transfers the data from the module into a 8-word file beginning at word address 070 through 077. At the completion of the read instruction, done bit 110/17 is set to 1 and the processor continues with other instructions in the ladder program.

The programming format for both a block transfer read and write instruction are the same and have the following meanings:

1. *Data address* is the address in the accumulated area of the data table.

2. *Module address* is determined by the rack (R), module group (G), and slot (S) numbers.

3. *Block length* is the number of words to be transferred.

4. *File* is the address of the first word of the file.

5. *Enable bit* (EN) is automatically entered based on the module address by the processor. It is set ON when the rung containing the instruction is true.

6. *Done bit* (DN) is automatically entered based on the module address. It remains ON for one processor scan following a successful transfer.

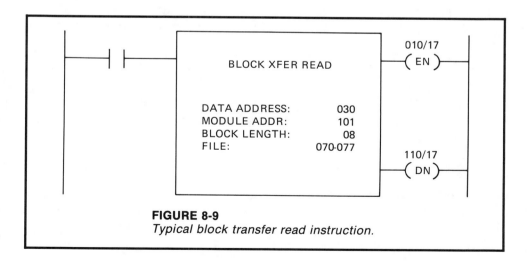

FIGURE 8-9
Typical block transfer read instruction.

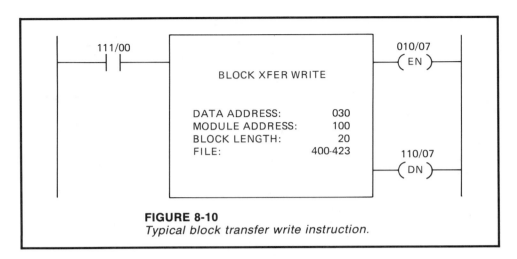

FIGURE 8-10
Typical block transfer write instruction.

An example program containing a block transfer write instruction is shown in Figure 8-10. The following parameters have been entered into the instruction using a programming panel: data address 030, module address 100, block length 20, and file 400.

In Figure 8-10, the module is located in rack 1, module group 0, and slot 0. The control and status bytes corresponding to the address of the module are at word locations 010 and 110 (lower bytes), respectively. The address of read enable bit is 010/07, and the address of the done bit is 110/07 for the read instruction.

During the program scan when bit 111/00 is true, the write instruction is enabled and the enable bit is set to 1. In the next scan of the processor, the control byte at word 010 is sent to the intelligent module. This control byte contains the instructions that inform the intelligent module that data will be sent. The module responds that it is ready to receive data. The processor then searches the data table and finds module address 100 in word address 030 and the file address, 400, in word 130.

The processor then transfers the data in memory addresses 400 through 423 to the intelligent module. At the completion of the write instruction, done bit 110/07 is set to 1 and the processor continues with other instructions in program memory.

EXAMPLE 8-4

Problem: Design a software program to read data every 1/2 second from an 8-channel analog input module in rack 2, module group 3, slot 1.

Solution: We can use a free-running timer at location 40 to execute the read command each 1/2 second as shown.

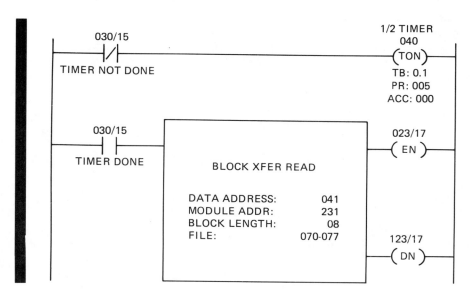

Shift Register Instructions The common file shift instructions encountered are: (1) shift file up, (2) shift file down, (3) FIFO load, and (4) FIFO unload. The first two shift register instructions are used to construct synchronous word shift registers normally from 1 to 999 words long (see Figure 8-11). In these output instructions, on a false-true transition of a rung decision, the data from

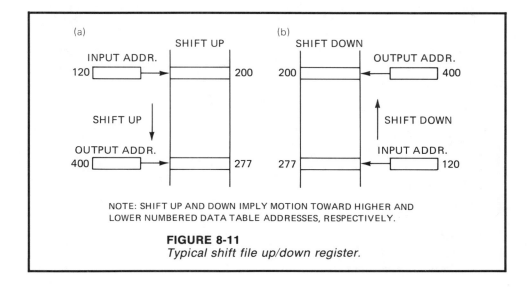

NOTE: SHIFT UP AND DOWN IMPLY MOTION TOWARD HIGHER AND LOWER NUMBERED DATA TABLE ADDRESSES, RESPECTIVELY.

FIGURE 8-11
Typical shift file up/down register.

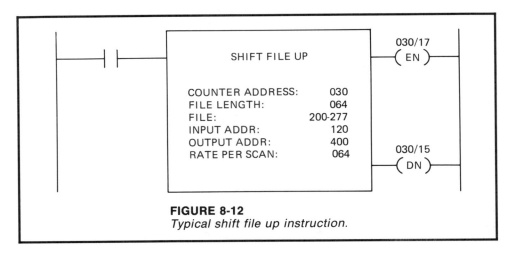

FIGURE 8-12
Typical shift file up instruction.

the input word will be shifted into the file, and the data in the last/first word of the file will be shifted up/down into the output word.

Shift File Up Instruction The shift file up instruction is used as a synchronous word shift register. When the instruction rung goes true, the data from a specified input word is shifted into the first word of the file (Figure 8-11 (a)), the data in the file is shifted up one word, and the information of the last word in the file is shifted into the specified output word. A typical shift file up instruction is shown in Figure 8-12.

In Figure 8-12, the *counter address* 030 is the address of the instruction. The *file length* is 064, so the file is 64 decimal words long. The *file* lists memory locations 200 through 277 for the data operated on by the instruction. *Input address* 120 is the memory address where data is entered. *Output address* 400 is the memory address of the output word. The rate *per scan* is 064, so all 64 data words in the file are operated upon each scan.

An example problem will illustrate the utility of a shift file up instruction.

EXAMPLE 8-5

Problem: Use a block transfer read instruction and a shift file up instruction to save the last five readings from a level transmitter. Assume the level transmitter is connected to the second channel on an 8-channel analog input module in rack 1, module group 4, and slot 0.

Solution: The block transfer read instruction places the data from the second channel into word 401. Then this word is transferred at the end of a read cycle into a holding file starting at location 500 as shown.

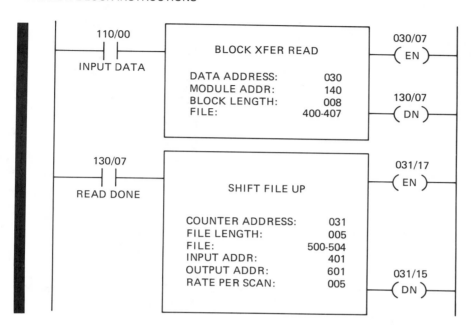

Shift File Down Instruction The shift file down instruction is used as a synchronous word shift register. When the instruction rung goes true, the information from a specified input word is shifted into the last word of the file (Figure 8-11 (b)), the data in the file is shifted down one word, and the information of the first word in the file is shifted into the specified output word.

A typical shift file down instruction is shown in Figure 8-13. The meaning of each term in the instruction is the same as the shift file up instruction.

FIFO Load and FIFO Unload Instructions The FIFO load and FIFO unload instructions are used together to make an asynchronous word shift register (see Figure 8-14). The FIFO load instruction transfers information from a specified input address into the file, on a false-true transition of a rung decision. The FIFO unload instruction transfers data from a file into a specified output address on a false-true transition of a rung decision.

Load and unload pointers keep track of the load and unload addresses in the FIFO file so that words are taken from the file in the entered—thus the acronym FIFO (first in, first out). These pointers are controlled by the processor as is the data in the FIFO file. The load and unload pointers can load and unload data at any point in the file. The body of the file will float between these boundaries. The user should not expect to find any particular data entry at any specific data table location in the FIFO file.

Only the FIFO input and output addresses are pertinent to the FIFO operation, and they are the only words that should be manipulated by, or examined by, the user. Typical FIFO load and FIFO unload instructions are

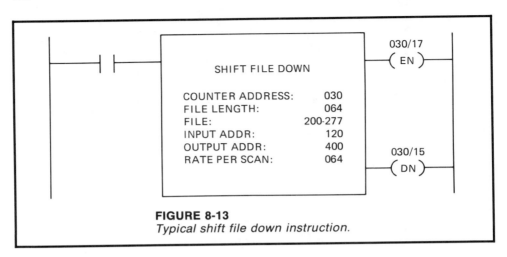

FIGURE 8-13
Typical shift file down instruction.

shown in Figures 8-15a and b. It is important to note that the FIFO load and FIFO unload instructions must have the same counter address, the same size, and the same file address.

The *counter address* is the memory address of the instruction in the accumulate value area of the data table. The *FIFO size* is the maximum number of words that the stack can hold. The *number in file* is the current number of words in the FIFO stack. The *file* gives the starting and ending address of the stack. The *input address* is the memory address of the input word outside the stack. The *input data* is the current information in the input address. The *output address* is the memory address of the output word outside the stack. The *output data* is the current data in the output address.

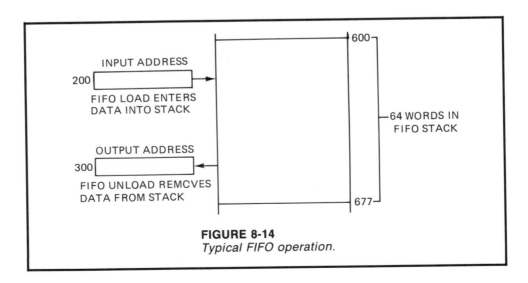

FIGURE 8-14
Typical FIFO operation.

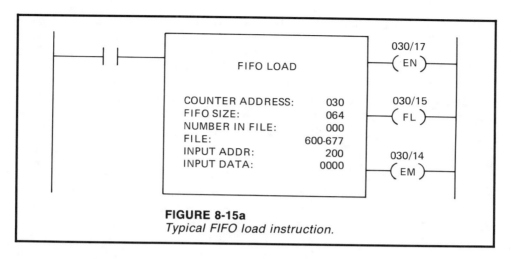

FIGURE 8-15a
Typical FIFO load instruction.

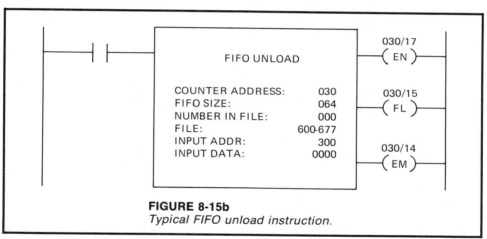

FIGURE 8-15b
Typical FIFO unload instruction.

The status bits are enable (EN), full (FL), and empty (EM). If the FIFO file is full and FL bit 15 is set, any data that is loaded will be lost. Conversely, if the FIFO stack is empty and EM bit 14 is set, no data can be unloaded. The programmer should use the full and empty flags to ensure that the data being loaded or unloaded to or from the file is not lost or invalid.

Sequencer Instructions   The sequencer instructions are the most powerful function block instructions available in programmable controller systems. They normally operate on up to 4 words (64 bits) at a time. There are three common sequence instructions: (1) sequencer output, (2) sequencer input, and (3) sequencer load.

The sequencer instructions are used to transfer data from the data table to discrete output modules for the control of sequential process operations or sequential batch operations (sequencer output). They are also used to compare I/O word data with data stored in tables so that process operating conditions can be examined for control and diagnostic purposes (sequencer input). The instruction is also used to transfer I/O word data into the data table (sequencer load).

Sequencer instructions are similar to file instructions, but there are some important differences. Both are function block instructions that contain a counter and a file. The instructions require the entry of more than one address, and each has a corresponding data monitor display for monitoring, loading, and editing file data.

File instructions operate on files that are one word or 16 bits wide. In contrast, sequencer instructions process files that are normally up to four words or 64 bits wide. A sequencer file can be represented graphically by a sequencer table or state table. It is sometimes called a state table because each row can represent the "state" of the process for a given fixed time or step in a given batch process. In most PC systems, the length or number of rows (steps) in a state table is limited to 999, and the width is limited to four words (64 bits) as shown in Figure 8-16.

Sequencer Output Instructions

The sequencer output instruction functions in the following manner, when the rung containing the sequencer output instruction goes from false to true, the counter increments to the next step in the sequencer table. The data found in that step is output to the output word address specified in the sequencer instruction. The output word addresses do not need to be consecutive. The sequencer output instruction normally operates whenever the rung is true. This means that once the rung is true, the counter is incremented to the next step and the data in that step will be output every processor scan that the rung remains ON. However, the instruction will not advance to the next step until there is a false-to-true transition of the rung.

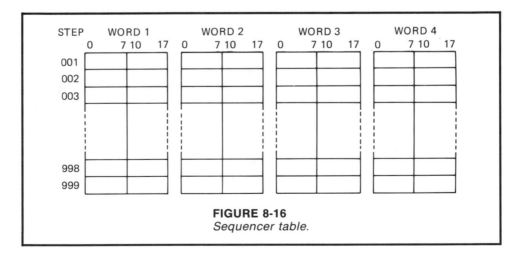

FIGURE 8-16
Sequencer table.

The done (DN) is set when the last step of the state table is output. The next false-to-true transition of the rung will start the operation again at step 1. The instruction counter is normally set to zero for start-up purposes if it is desirable to start at step 1 when the instruction is enabled for the first time.

The sequencer output instruction has mask capability. A mask is a means of selectively screening out data. It allows unused bits of output words specified in the instruction to be used for other purposes. The sequencer output instruction operates through the output words specified in the instruction. The number of output bits required for the batch operations can be less than 16, 32, 48, or 64, corresponding to 1, 2, 3, or 4 word sequencer instruction. When this is true, masking allows the unused output terminals of a module selected in the instruction to be used for other purposes.

A zero in a mask bit location prevents the instruction from affecting the data in the corresponding bit location. A 1 in a bit mask location allows the corresponding bit to be used. When all the output data bits are used by the sequencer instruction, a mask of all ones must be used.

An example rung containing the sequencer output instruction is shown in Figure 8-17.

In the example of Figure 8-17, when input bit 111/00 closes, the sequencer output instruction increments to step 008 and controls the 32 outputs in output words 011 and 013. The control of the output module will be in accordance with the data stored in step 008 of the sequencer table and the mask conditions programmed into words 211 and 212.

Sequencer Input Instruction

The sequencer input instruction compares process or machine input data with information stored in memory for equality. It is used alone or in a series and/

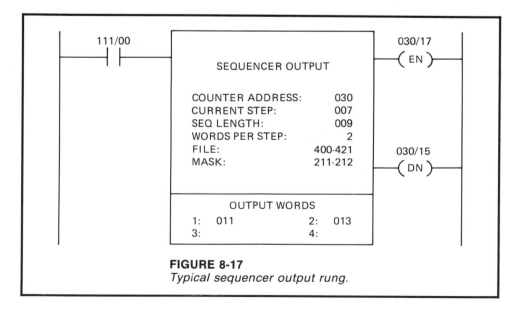

FIGURE 8-17
Typical sequencer output rung.

or parallel combination with other rung condition instructions to determine the status of an output.

This instruction normally compares the status of up to 64 input conditions with the data stored in a sequencer table, bit by bit and step by step. The instruction becomes logically true when the state of all input bits are equal to the state of all bits in the current step of the sequencer table.

The sequencer input instruction has a step counter that points to the step in the sequencer file being operated on. The counter is not controlled by the instruction. Its accumulated value is stepped by logic from another part of the control program.

The sequencer input instruction can be programmed in the same logic rung as a sequencer output instruction, and its step counter can be indexed by the sequencer output instruction. The step counters in both instructions have the same address. If programmed in the same rung, the sequencer input and output instructions will track each other through a controlled sequence of operations. The length of the sequence is equal to the number of steps in the sequencer table.

A typical sequencer input instruction is shown in Figure 8-18. In this instruction, up to four input word addresses can be specified, and each input word has a corresponding mask word. The mask is applied to the information at the input address when the bit comparisons are made. If the number of input bits used by the instruction is less than 16, 32, 48, or 64 for a 1, 2, 3, or 4 word sequencer instruction, the bits not used by the instruction should be masked. This allows unused input terminals of the specified input word to be used for other purposes in the control program.

A typical application for this instruction is to compare the status of control

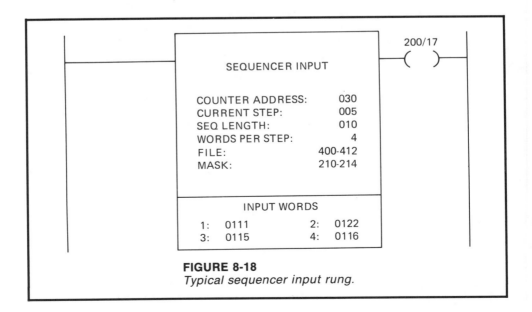

FIGURE 8-18
Typical sequencer input rung.

valves on a process stored in memory to their position value (open or closed) after a control step in a process.

In the example rung in Figure 8-18, when the state of all 64 inputs corresponding to the input words 0111, 0112, 0115, and 0116 (except the inputs that are masked) is equal to the state of all 64 bits in step 5 of the sequencer table, the instruction output bit 200/17 will be turned on.

Sequencer Load Instruction

The sequencer load instruction is used to load information into memory locations such as files or sequencer tables. The instruction receives information from up to 4 independent memory locations specified in the instruction. The input words can represent input, output, or storage words, and the load word addresses do not have to be consecutive.

The instruction loads words into a sequencer table or other parts of memory, one step at a time, at the location determined by the step counter. The instruction is started by a false-to-true transition of the rung input condition. Information from the load words will be stored in memory during the processor scan that the instruction has initiated. If the rung stays true for the subsequent scans, the instruction will not repeat. The instruction will step to the next position only on the next false-to-true transition of the rung input bit.

An example sequencer load instruction is shown in Figure 8-19. In this example, when input 111/00 is true, the sequencer load instruction will increment to step 008. The data from input words 0112, 0113, 0120, and 0300 are loaded into step 008 of the sequencer table in one processor scan. A new set of data will be loaded when input switch 111/00 opens, then closes again.

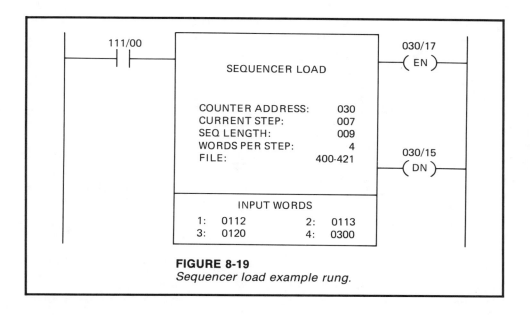

FIGURE 8-19
Sequencer load example rung.

Sequential Function Chart Language
Any software design is essentially a modular process. The design starts by breaking the problem down into understandable modules, with flow between modules or processes generally determined by some decision. A useful method employed by programmers is the flowchart, which is simply a tool for describing a sequence of events or processes. The common symbols used in flowcharts are shown in Figure 8-20. The two most important symbols are a block for a single process and a diamond for a decision. The number of flowchart pages, as with block diagrams, is strictly a function of the complexity of the program.

The symbols for on-page and off-page connections are used to reduce the number of interconnecting lines on a diagram or diagrams, which improves the clarity of the flowchart.

Since flowcharting is a standard procedure used by programmers to develop and document software, some programmable controller vendors have designed their machines to be programmed directly in a similar language called sequential function chart.

Flowcharting languages offer several advantages, such as improved organization, modularity, and readability, over free-form programming. A small change in one section of the program generally does not cause problems elsewhere. This leads to faster program debugging, easier modifications, and improved software reliability.

In most programmble controllers that use sequential function charts, each process or function block is written in ladder logic so that the system designer is able to use a top-down approach to directly transform a control problem into a programming solution. First, the control logic is subdivided into smaller and smaller blocks until each block or process step is directly represented in the function chart. Then each step is programmed using standard ladder logic.

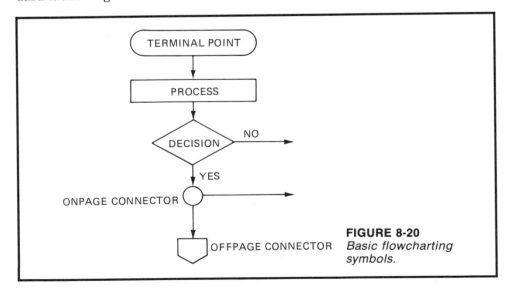

FIGURE 8-20 *Basic flowcharting symbols.*

An advantage of this method is that the processor scans only the active steps and transitions (decisions) of the sequential function chart that correspond to the particular stages of the process or machine operation that are currently active. All other steps in the control program are not scanned. This greatly reduces the effective program scan time.

Another advantage of sequential function chart programming is that it is self-documenting. As the programmer builds the software on a programming terminal, it is displayed on a CRT (cathode ray terminal) for monitoring and debugging, and it can be easily printed out for future reference.

An example sequential function chart program is shown in Figure 8-21 to explain the symbols used in a typical program. The top of the program contains a *start* block to define the beginning of the program. The next block is the *initial step*, where the programmable controller starts function chart execution and returns to this step from the end of program unless directed otherwise by the program logic. This block is identified by a double-sided box.

The *step* block is the function chart's basic unit and contains ladder logic for each independent stage of the process or machine operation. It is identified by a single-sided box.

The *transition* is the logic condition that the processor checks after com-

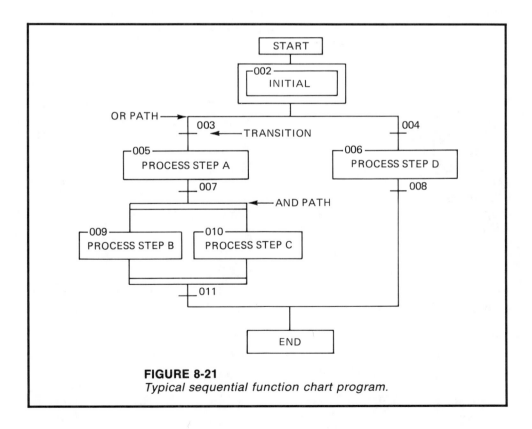

FIGURE 8-21
Typical sequential function chart program.

pleting the active step. When the transition logic is true, the step preceding the transition is disabled, and the step following it becomes active. The transition is normally a single logic rung, identified by a short horizontal line below its corresponding step (see Figure 8-21).

The *OR path* is identified by a single horizontal line at the beginning and end of a logic zone. The processor selects one of several parallel paths depending on which transition goes true first. Normally, the number of parallel paths is limited to seven.

The *AND path* is identified by a horizontal double line at the beginning and end of a zone. The processor can normally execute up to seven paths at the same time.

In sequential function chart programs, steps and transitions are arranged in series and parallel paths, and they are numbered with the file numbers that contain their ladder logic. The programmable controller scans the logic of a step repeatedly until its transition logic goes true. Then the program scan moves to the next step or steps, and the previously active step is turned off.

There are three basic rules for standard sequential function chart programs. The first is that the *initial* step is always activated at start-up. When restarting from the beginning of the chart and on subsequent passes through the flowchart program, the programmer does have the option of restarting from the beginning of the last active step(s), following or changing the programmable controller's mode from test or run to program.

The second rule is that a transition is tested after its associated step, and operations pass from one step to the next through a transition when the transition goes true.

The third rule is that after a true transition, the processor scans the step once more to reset all timer instructions and then executes the next step. This extra processor scan is called postscan. It is important to note that the processor never postscans a transition file, so timers probably should not be used in transition files.

The processor scans a sequential function chart program from left to right and top to bottom. When the sequential function chart program scan encounters active parallel steps, it executes the ladder logic in the leftmost step first, then moves to the ladder logic in the next parallel step across the screen from left to right.

Grafcet Graphical Function Chart Programming[6]   One example of sequential function chart programming for PCs is a French national standard called Grafcet. Grafcet is a graphical function chart programming language adapted into a programmable controller language by Telemechanique.

Grafcet is intended as a universal sequential control language to be used on all types of control systems: electronic, pneumatic, hydraulic, or mechanical. It describes all control functions and produces a structured, logical approach to control system design and analysis. It also has the advantage of being self-documenting and is easy to follow for simple control problems.

Ladder logic programming is also self-documenting and easy to read, but it becomes cumbersome and difficult to follow if the control problem is time dependent and sequential. To illustrate the problem, consider a simplified example of a semiautomatic punch, shown in Figure 8-22. We assume the punch starts in the raised position or top position. When the operator depresses the start push button, the punch is lowered and it pierces the metal part at the lowest or bottom position. The cycle is completed when the control system raises the punch back to the top position.

The PC ladder program, shown in Figure 8-23, not only must take into consideration that depressing the start button turns on the punch drive motor, but that the operator will also release the start button. The program must also

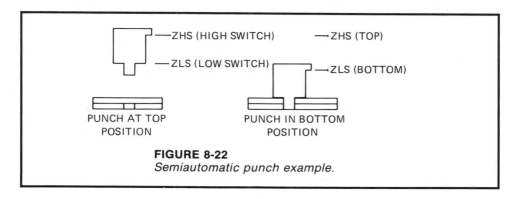

FIGURE 8-22
Semiautomatic punch example.

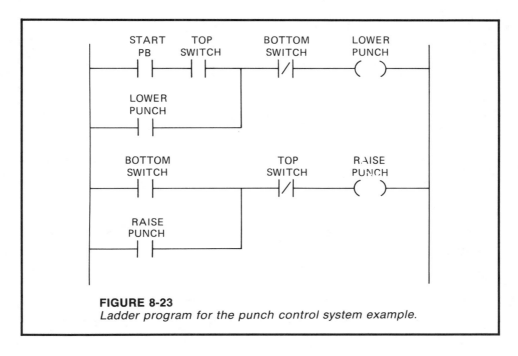

FIGURE 8-23
Ladder program for the punch control system example.

take into consideration that the operator might depress the start PB when the punch is in motion. All these additional facts, which have no bearing on the action of the punch, complicate the description of the control function.

On the other hand, the sequential function chart method represents control algorithms in a sequential manner, showing the actions that take place in each step and indicating the conditions that must be met in order to advance to the next step. The flowchart for the semiautomatic punch example is shown in Figure 8-24.

The Grafcet diagram uses a small set of programming symbols as shown in Figure 8-25. Steps (actions) are blocks and transitions, shown using horizontal lines. Each step is followed by a transition and each transition symbol must be followed by a process. The "Initial" step symbol is a square with double lines. The process "Steps" are represented by numbered squares that can contain a mnemonic indicating the function. The "Actions" are described in the rectangles to the right of the step. This provides for program documentation that is performed as the system is being programmed. The "Transitions" are shown using short horizontal lines in the flowchart. The "Conditions" asso-

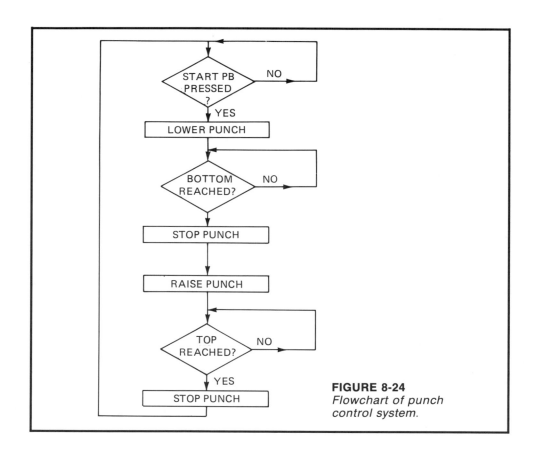

FIGURE 8-24
Flowchart of punch control system.

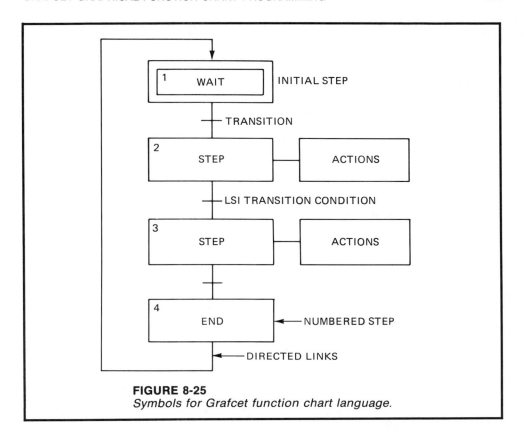

FIGURE 8-25
Symbols for Grafcet function chart language.

ciated with a transition are listed to the right of the transition. The "Directed Links" indicate sequential process flow from the top to the bottom of the chart.

The main advantage of a sequential function chart is that the process or machine sequence can be clearly seen in the chart. However, a single sequential function chart program has a problem: In most complex control applications the control computer cannot simply go into a loop waiting for an event to occur. It must be able to monitor for interrupt conditions, such as emergency stops. Also, in most control systems several sequences must be controlled simultaneously. All these requirements force the programmer to add interrupt flags and jump statements that complicate the flowchart program and destroy the clarity of the program.

The Grafcet function chart language has provisions for simultaneous control sequences. This is a major advantage over standard flowchart languages that allow only one action or one interrogation at a time.

In the Grafcet flowchart language representation of the simple semiautomatic punch control system shown in Figure 8-26, the system has three states or steps. The first state is the punch at rest in the raised position; the second step is the punch descending; the final step is the punch ascending.

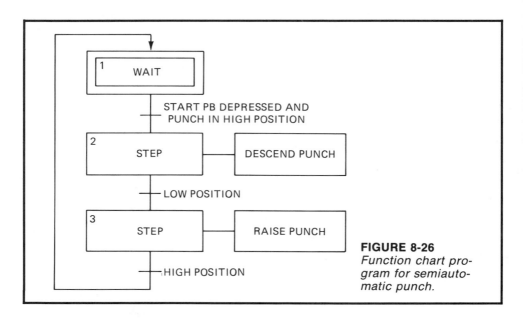

FIGURE 8-26
Function chart program for semiautomatic punch.

When the punch and associated control system are activated, the programmable controller places the system in the Wait state (step 1) and waits for the first transition logic to be satisfied. If the start PB is depressed AND the punch is in the top position, the first transition is true and step 2 is activated, so the punch moves down. At the same time, the programmable controller turns OFF the first state.

When the punch activates the bottom limit switch, the second transition is true, and the punch raises. Finally, when the punch reaches the top, the final transition is set and the program returns to the Wait state.

This simple sequential function chart programming example shows advantages of a graphical programming language over other programming methods. The flowchart program in Figure 8-26 is much easier to follow than the ladder logic program for the same application shown in Figure 8-23.

EXERCISES

8.1 Use data transfer file instructions to write a program to transfer process batch numbers from a 3-digit BCD input module to 20 consecutive memory locations in a programmable controller. Assume the operator uses a normally open push button at location 111/16 to enter each BCD number. Use an LED indicator at output location 050 to display the batch number. Use an LED indicator at output location 050 to display the batch number (1 through 20) and an another LED to display the data at the batch step selected.

8.2 Write a program to transfer 20 words of data starting at memory word 1000 to a block of memory starting at word 800 if input bit 100/13 is true.

8.3 Write a software program to transfer 10 words of process data from consecutive memory locations starting at word 400 to a panel-mounted LED display. Assume the operator uses a control panel push button at address 100/10 to request process data. Use a second LED display to indicate the memory location of the data to the operator. The LED displays are connected to the digital output modules at memory locations 700 and 701.

8.4 Design a software program to read data every 0.02 second from an 8-channel analog input module in rack 1, module group 5, slot 0. Send the same data to an 8-channel analog output module in rack 0, module group 7, slot 0 right after the data read instruction has been completed.

8.5 Use file shift file instructions to write a program to transfer process set point numbers from a 3-digit BCD input module to 8 consecutive memory locations in a programmable controller. Assume the operator uses a normaly open push button at location 100/00 to enter each BCD number.

8.6 Write a program to store the last ten temperature readings from the seventh input channel of an 8-channel analog input module located in rack 1, module group 2, and slot 1.

BIBLIOGRAPHY

1. Jones, C. T., and Bryan, L. A., *Programmable Controllers, Concepts and Applications,* International Programmable Controls, Inc., First Edition, 1983.

2. *Modicon* 584; *Programmable Controllers, User's Manual,* Gould Modicon Division, January 1982.

3. *PLC-2/30 Programmable Controller: Programming and Operations Manual,* Allen-Bradley, Publication 1772-6.8.3, 1984.

4. Hughes, T. A., *Basics of Measurement and Control,* Instrument Society of America, 1988.

5. Gilbert, R. A., and Llewellyn, J. A., *Programmable Controllers–Practices and Concepts,* Industrial Training Corporation, 1985.

6. Lloyd, M., "Grafcet Graphical Functional Chart Programming for Programmable Controllers," *Measurement & Control Magazine,* September, 1987 edition.

9

Data Communication Systems

Introduction   The purpose of communications is to transfer information from one point to another or from one system to another. In process control this information is called process data or, simply, data.

An understanding of data communications is essential for the proper application of programmable controllers to process control and data collection. This chapter will provide a basic understanding of data communications terminology and concepts and their application to programmable controller systems.

Basic Communications   Data is transmitted through two types of signals: baseband and broadband. In a baseband system, a data transmission consists of a range of signals sent on the transmission medium without being translated in frequency. A telephone call is an example of a baseband transmission. A human voice signal in the 300- to 3000-Hz range is transmitted over the phone line in the 300- to 3000-Hz range. In a baseband system there is only one set of signals on the medium at a time.

A broadband transmission consists of multiple sets of signals. Each set of signals is converted to a frequency range that will not interfere with other signals on the medium. Cable television is an example of broadband transmission.

Three basic components are required in any data communication system: (1) the *transmitter* to generate the information, (2) the *medium* to carry the data, and (3) the *receiver* to detect the data, as shown in Figure 9-1.

The medium can be divided into more than one channel. A channel is defined as a path through the medium that can carry information in only one direction at a time.

Figure 9-2 shows an example of a four-channel medium. Probably the

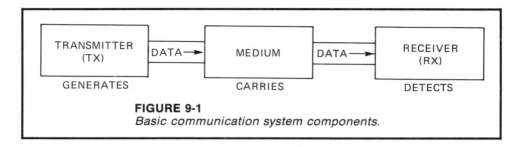

FIGURE 9-1
Basic communication system components.

most common example of communication channels is television stations on home TVs.

Transmission media fall into three general categories: twisted pair, coaxial cable, and fiber optic cable. Twisted pair consists of two electrical conductors, each covered with insulation. The two wires are twisted together as shown in Figure 9-3 to ensure that they are both equally exposed to electrical interference signals in the environment. Since the wires are carrying current in opposite directions, the electrical interference will tend to cancel out in the cable. Twisted pair is the most common cable used in programmable controller systems. It is the least expensive transmission medium and provides adequate electromagnetic interference (emi) immunity.

A coaxial cable consists of an electrical conductor surrounded first by insulating material and then by a tube-shaped metal braid conductor. In most cases, the entire cable is covered by an insulator, as shown in Figure 9-4. The round conductor in the center of the cable and the circular outer tube conductor are coaxial in that they share the same central axis.

Coaxial cables are used in both baseband and broadband communications systems. A variety of cable types are used in process control applications, ranging from flexible types similar to coaxial cables used on home TV sets to heavy, rigid cables that require special installation procedures.

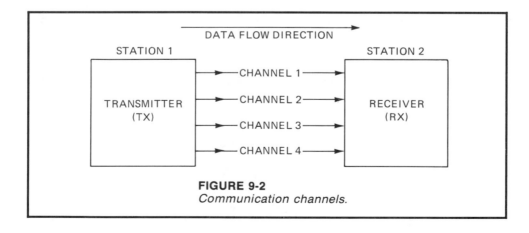

FIGURE 9-2
Communication channels.

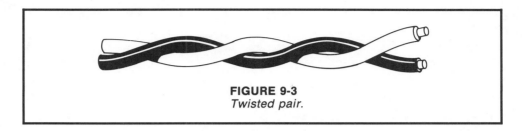

FIGURE 9-3
Twisted pair.

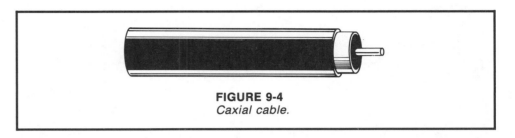

FIGURE 9-4
Caxial cable.

Coaxial cables are generally used in process automation applications where long communications runs of over 1000 feet are involved and improved emi immunity is required. They find very extensive use in plant-wide communication networks.

The fiber optic cable consists of fine glass or plastic fibers, as shown in Figure 9-5. At one end, electrical pulses are converted into light by a photo diode and sent down the fiber optic cable. At the other end of the cable, a light detector converts the light pulses back to electrical signals. The light signals can travel only in one direction, so two-way transmission requires two separate fiber cables. A fiber optic cable is normally the same size as a twisted pair cable, and it is immune to electrical noise.

The cost of the optical fiber cables is about the same as coaxial cables. However, the use of fiber optics in programmable controller applications has been limited by the high cost of connectors and lack of industrial standards.

Communication can also be described by the number of channels used to effect the information flow. The three common methods of data transmission are simplex, half duplex, and full duplex.

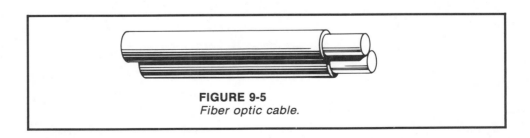

FIGURE 9-5
Fiber optic cable.

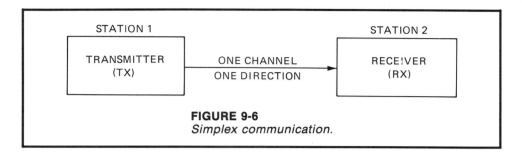

FIGURE 9-6
Simplex communication.

Simplex Communications   In simplex communications, a single channel is used and communication is only one way (i.e., from transmitter to receiver), so the receiver can never respond (see Figure 9-6).

In simplex communication, it is not possible to send error or control signals from the receiver station, because the transmitter (TX) and the receiver (RX) are each dedicated to performing one function. A typical example of simplex communications in process control would be a weigh scale in the field sending data to a computer in a central control room.

Half Duplex Communications   Two-way communication allows the receiver to verify that the data was received. One kind of two-way communication is called half duplex. In half duplex, a single channel is used and communication

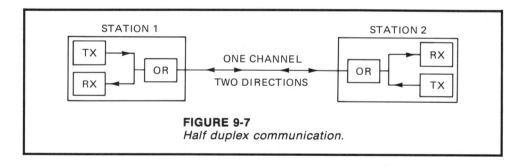

FIGURE 9-7
Half duplex communication.

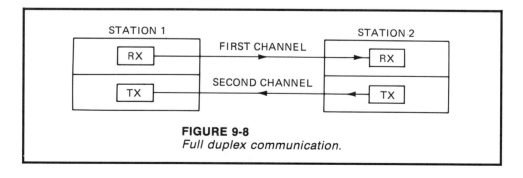

FIGURE 9-8
Full duplex communication.

is two-way, but communication can occur only in one direction at a time (see Figure 9-7). In this configuration, the receiver and transmitter alternate functions, so communication occurs in one direction at a time on the single channel.

Full Duplex Communications   Two-way communications where data can flow in both directions at the same time is called full duplex communications. In this case two channels are required, so information can flow from both directions simultaneously, as shown in Figure 9-8.

Transmission Methods   Transmission involves moving data between a receiver and a transmitter. The two methods of transmission are *parallel* and *serial*.

Parallel Transmission   In parallel transmission, all bits are transmitted at the same time, and each bit of information requires a unique channel. For example, if we transmit an eight-bit ASCII character, eight channels are required. Figure 9-9 demonstrates the transmission of the ASCII character G using even parity. The term "parallel" refers to the position of the bits of the character and the fact that the characters are transferred one after the other. The use of several channels results in a high data transfer rate.

There is a problem with bit synchronization delay in parallel data transfer. Synchronization is the process of making two or more activities happen at the same rate and time. If we send a string of characters on a long parallel cable, the difference in impedances in the cable wires can cause bit synchronization decay, as shown in Figure 9-10.

To overcome this problem, the distance between the receiver and the transmitter must be kept short to avoid transfer errors caused by bit syn-

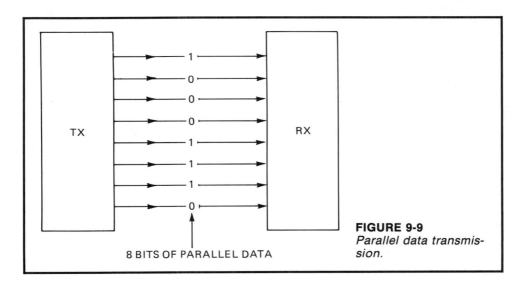

8 BITS OF PARALLEL DATA

FIGURE 9-9
Parallel data transmission.

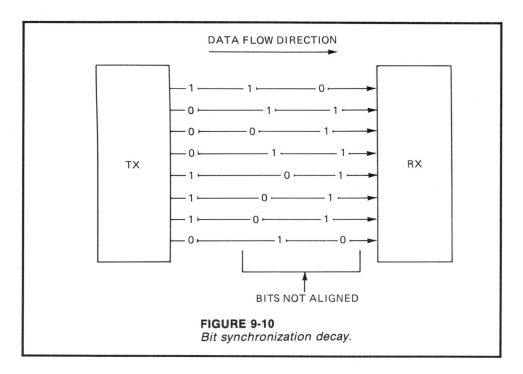

FIGURE 9-10
Bit synchronization decay.

chronization decay. For this reason, parallel transmission is used only in closely connected data systems where high speed is required. For example, data transfer in a microcomputer system is parallel transmission. However, in process control systems the data transfer between systems such as personal computers and programmable controllers is normally serial. For this reason, we will limit our discussion in the remaining parts of this chapter to serial data transmission.

Serial Transmission   In this section, we will discuss serial interfacing between microcomputer-based devices. Serial interfacing is probably the least complex electrical interface to implement because it requires only one signal wire to carry all data flow in one direction and only two wires for bi-directional flow. Serial interfaces do, however, require additional logic circuitry to convert between parallel and serial data because most computers process data in parallel. This extra circuitry is not required in parallel transmission of information. Because serial links are low in cost and easy to install, they have been standardized into a few widely used protocols (transmission control standards), where several low-cost integrated circuit (IC) interface chips are available.

In serial transmission, the bits of the encoded character are transmitted one after another in a single channel (see Figure 9-11). The transmission takes

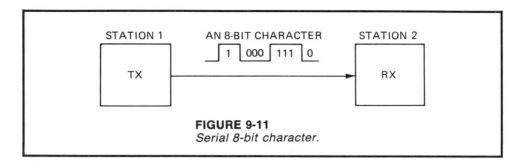

FIGURE 9-11
Serial 8-bit character.

the form of a bit stream that the receiver must assemble into characters (normally 8 bits) using specially designed ICs.

The main advantages of serial communications are lower cost and the elimination of byte synchronization problems.

Controlling the speed of transmission is critical in serial data transfer. A common measurement unit used to describe serial data speed is *bits per sec* (bps).

The process of bit synchronization is required for serial data transfer. A timing circuit at the transmitter places transmitted bits on the channel at fixed intervals determined by the selected communication rates. Some typical communication rates used are 110, 300, 600, 1200, 2400, 4800, and 9600 bps.

The transmitter must send bits at the same rate the receiver is set to accept bits. For example, if a personal computer is transmitting serial data at 9600 bps to a programmable controller system, the communications module on the programmable controller must be set at a communications rate of 9600 bps.

Character synchronization is used to determine which eight consecutive bits in a data stream represent a character. The receiver must recognize the first data bit and count bits until the character or byte is complete. Various forms of character synchronization are used in data communications. Sometimes each byte is framed with special bits to indicate start and stop.

Signal Multiplexing
A single line can be used to carry more than one signal by using signal multiplexing. To understand multiplexing, we must first define the term "bandwidth". Bandwidth describes the signal-carrying capability of a communications channel or line. Bandwidth is defined as the difference in cycles per second between the lowest and highest frequencies a channel can handle without a certain amount of signal loss.

Multiplexing allows one line to serve more than one receiver by creating slots or divisions in the line. The equipment used to obtain multiplexing is called a multiplexer or MUX (see Figure 9-12).

The commonly used signal multiplexing methods are frequency division, time division, and statistical.

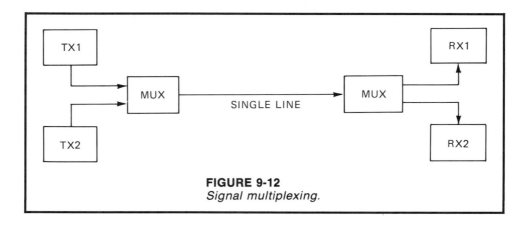

FIGURE 9-12
Signal multiplexing.

Frequency Division Multiplexing   In frequency division multiplexing each channel is assigned its own individual frequency range. Frequency division is normally used to combine a group of low-speed sources onto a single voice line. A typical application of frequency division communications is shown in Figure 9-13.

Time Division Multiplexing   Time division multiplexing uses time periods to assign channel space and allot the available bandwith. In Figure 9-14, two time periods (Period 1 and Period 2) are used to assign each channel its own individual time slot in the data stream. In this type of multiplexing, no other channel can use the time slot, so there is some wasted bandwidth when a station is not transmitting data in its time slot. However, one transmission line can support many data streams. It is normally used where there is a need to combine a number of relatively low-speed data transmissions onto a single high-speed line.

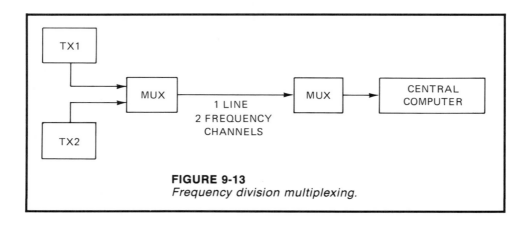

FIGURE 9-13
Frequency division multiplexing.

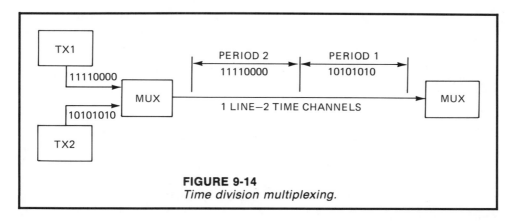

FIGURE 9-14
Time division multiplexing.

Statistical Multiplexing Statistical multiplexing is an enhancement of time division multiplexing designed to reduce the waste bandwidth when data is being transmitted by a station. This enhancement process is sometimes called data concentration. Statistical multiplexing is generally used where a large number of data entry terminals require only a brief or occasional data transfer. If the use of the terminals becomes intensive, the operators will experience long time delays, so this method must be applied with caution.

Error Control and Checking Error control and checking is the process used to detect any discrepancy between transmitted data and received data in a communication system. Errors detected at the receiver are either corrected or retransmitted. The common error control methods used in programmable controller systems are (1) echo check (Echoplex), (2) vertical redundancy check (VRC), (3) longitudinal redundancy check (LRC), and (4) cyclical redundancy check (CRC).

Echo Check Echo check is used in two-way communications systems (Figure 9-15) to check the accuracy of transmitted data.

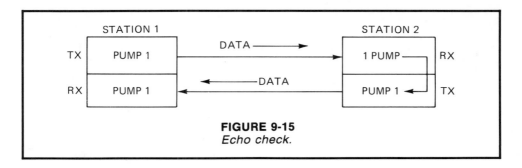

FIGURE 9-15
Echo check.

The receiver (RX) in station 2 sends every character received back to the originating terminal (station 1). Station 1 compares the echoed data with the information it sent. The comparison may be performed by the station operator or by electronic circuitry in the transmitter. If a transmission error is detected, the information is retransmitted.

Vertical Redundancy Check (VRC)   In the vertical redundancy check process, parity bit checking is used to detect the change of a single bit. In this method, a single bit is added to a character string to create either an odd or an even number of 1 bits. Parity bits are often called "redundant" because they can be removed without loss of data. If even parity is used, a 1 bit is added to a character string to make the total of 1 bits even. For example, the ASCII letter "A" is given by 0000001; if even parity is used, a 1 bit is added to the binary string before transmission (i.e., A = 10000001).

In even parity VRC transmission, the receiver would detect an error when it receives a character that contains an odd number of 1 bits. In this method, after a parity error is detected, the receiver would request retransmission of the data.

An example problem will help to explain the concept of parity.

EXAMPLE 9-1

Problem: Generate the ASCII code with even parity for the word STOP.

Solution: First using the ASCII table in Chapter 2 we obtain the binary code for each character as follows:

Bits	7 6 5 4 3 2 1	Character
	1 0 1 0 0 1 0	S
	1 0 1 0 1 0 0	T
	1 0 0 1 1 1 1	O
	1 0 1 0 0 0 0	P

Next we count the number of 1 bits for each character. If the number is even, we set the parity (P) bit to 0 to maintain an even number of 1 bits in the binary string. On the other hand, if the number of 1 bits is odd, we set the parity bit to 1 to obtain an even number of 1 bits in each string as shown.

Bits	P 7 6 5 4 3 2 1	Character
	1 1 0 1 0 0 1 0	S
	1 1 0 1 0 1 0 0	T
	1 1 0 0 1 1 1 1	O
	0 1 0 1 0 0 0 0	P

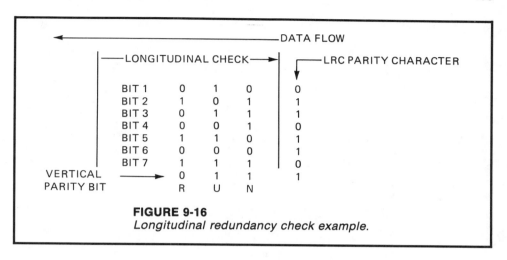

FIGURE 9-16
Longitudinal redundancy check example.

Longitudinal Redundancy Check (LRC)   Longitudinal redundancy check (LRC), or block check character (BCC), is a procedure that checks an entire horizontal line within a block of data for odd or even parity. This process is used in combination with VRC. While VRC can detect a single error, the only way to obtain corrected data is retransmission. When LRC is used with VRC, any single bit error in an entire data block is not only detected but can also be corrected at the transmitter without a retransmission. This increases the overall data transmission speed for a system.

To illustrate the longitudinal redundancy check process, let's assume we are transmitting the data word RUN in ASCII code using odd parity as shown in Figure 9-16.

Note that in Figure 9-16, an LRC odd parity character is sent by the transmitter. This character is generated by adding a parity bit at the end of the entire transmitted block so that an odd number of 1 bits is created for each longitudinal row of bits.

At the receiver, the LRC is calculated for the data bytes in the block, and it is compared with the transmitted LRC character. If they are not equal, the vertical parity bit reveals which byte is in error, and the LRC reveals which of the 8 bits is in error. Logic circuitry or software in the receiver is used to change the error bit to its opposite state, correcting the error. Only if there is more than one error in the block transfer must retransmission be requested by the receiver.

EXAMPLE 9-2

Problem: Calculate the LRC/VRC odd parity characters for the word PUMP.

Solution:

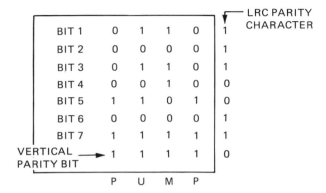

Cyclical Redundancy Check (CRC)   Cyclical redundancy check is a method for checking error on an entire data block. In this method, a check character is transmitted at the end of each block. The transmitter calculates the check character by dividing the binary number represented by the block of data by a constant, discarding the binary quotient and using the remainder as the block check character (BCC).

The receiver compares the transmitted block check character to its own BCC, calculated using the received block of data. If they are not equal, the receiver requests retransmission of the previous message block. If they are equal, the transmitted data is assumed to be error-free.

Cyclical redundancy check is a very accurate method for detecting data transmission errors. It is the most common method used in programmable controller-based systems.

Communication Protocols   Communications protocol is a set of
rules specifying the format and control of transmission between two communication devices. The two main functions of protocol are *handshaking* and *line discipline*. Handshaking determines whether a circuit is available and makes sure the circuit is ready to transfer data. Line discipline performs the following functions: (1) receive and transmit information, (2) error control procedures, (3) sequencing of the message blocks, and (4) error checking.

To illustrate the concept of communications protocol, a typical line discipline sequence between a transmitter (TX) and receiver (RX) is shown in Figure 9-17.

The two basic transmission methods used to obtain line discipline are:

1. *Asynchronous*: Successive data appear in the data channel at arbitrary times, with no specific clock control governing the time delays between information.

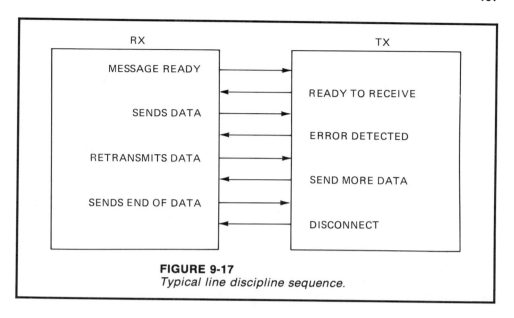

FIGURE 9-17
Typical line discipline sequence.

2. *Synchronous*: Each successive datum in a data stream is controlled by a master data clock and appears at a specific interval of time.

Most synchronous and asynchronous systems deliver serial data in 8-bit characters. The asynchronous systems treat each character as an individual message, and the characters appear in the data stream at arbitrary relative times. However, within each character the 8 bits are transmitted at a fixed predetermined clock rate, such as 110, 300, 600, or 1200 bps. Hence, an asynchronous communications system is actually synchronous within a character and asynchronous between characters.

A typical bit stream for an asynchronous transmission is shown in Figure 9-18. Line discipline is achieved by framing every character transmitted with start/stop bits. The start bit is a 0 or "space" and the stop bits are a 1 or "mark."

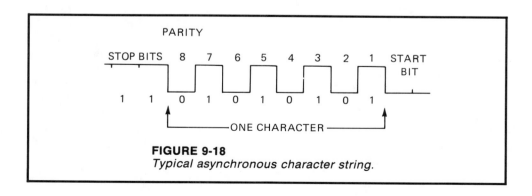

FIGURE 9-18
Typical asynchronous character string.

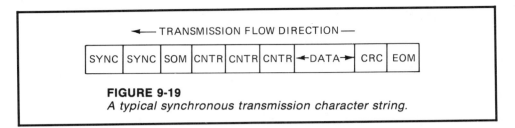

FIGURE 9-19
A typical synchronous transmission character string.

Asynchronous transmission operates at random speeds and is almost always used with half duplex protocols.

Synchronous transmission sends an entire message at one time and uses special character strings to achieve line discipline. It includes a predefined start and end sequence and normally operates at speeds of 2400 bps and higher. A typical sequence begins with two synchronous (syn) characters as shown in Figure 9-19.

In Figure 9-19, the start-of-message (SOM) character indicates the beginning of the data transfer, and the control character gives the receiver control information on the data transfer. The actual data can vary in length and is followed by error checking, which is normally CRC. The end-of-message (EOM) is used to signal the end of a data transmission.

The two transmission modes used for line discipline are half duplex and full duplex protocols. Half duplex protocol allows for communications in only one direction. In a 2-wire half duplex circuit, the stations must be able to switch automatically from transmit to receive. In a 4-wire full duplex circuit, each station has a dedicated receiver and transmitter circuit, but data cannot be transmitted and received at the same time. Full duplex protocol permits simultaneous transmissions in both directions. It falls back to half duplex operation if the communications link cannot support full duplex.

Serial Synchronous Transmission

Synchronous communications depend on the data and protocol control information being assembled into a structured and predefined package or format. The format tells the computer or other equipment what to do with the message and how to do it.

A message will contain one or more fields of data. Each synchronous transmission in a message block format will contain three major parts: (1) a header, (2) text, and (3) the trace or trail block, as shown in Figure 9-20.

The *header* starts the message package and provides the capability for synchronization between the transmitter and receiver. It normally contains characters that identify the transmitting or receiving stations and provides communications routing data. The *text* contains the data characters to be transferred and consists of a single- or multi-block message. The *trace* or *tail* characters signal the end of a transmission block.

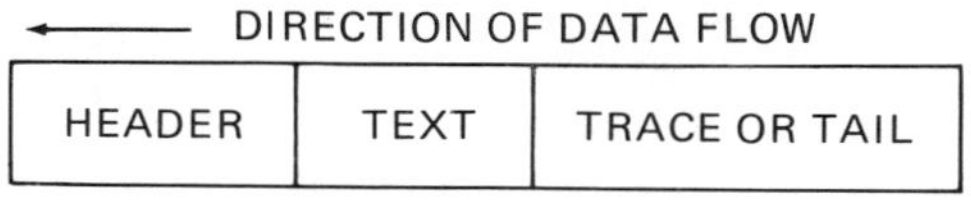

| ← DIRECTION OF DATA FLOW | | |
| HEADER | TEXT | TRACE OR TAIL |

FIGURE 9-20
Three elements of a typical synchronous transmission.

Serial Synchronous Protocols Different control and data formats are used by different protocols. A communications protocol can be defined as a fixed set of rules governing the format and control of inputs and outputs between two data transfer devices. In a standard data transfer system, a protocol governs the following: (1) line control, (2) framing, (3) error control, and (4) sequence control.

Line control is used to list the station that will transmit and the station that will receive in half duplex communications. It is also required in a multipoint circuit (a circuit connecting 3 or more points on a common line). The framing function normally determines which 8-bit groups are the characters and what groups of characters are the messages. Error control detects transmission errors using various redundancy checks and normally corrects faulty messages. Sequence control numbers the messages to eliminate duplication and avoid data loss. It also identifies any retransmitted messages.

There are basically three protocols in wide use today:

1. BISYNC (Binary Synchronous Communications), an older protocol used in IBM equipment

2. DDCMP (Digital Data Communication Message Protocol), a protocol used primarily by Digital Equipment Corporation (DEC)™ equipment

3. HDLC (High-level Data-Link Control), probably the most widely used protocol

Because BISYNC and HDLC protocols are the most widely used in programmable controller systems, we will limit our discussion to these two techniques. HDLC has evolved from two earlier standards, the ADCCP (advanced data communications control procedure) and the SDLC (synchronous data link control). Because of the close similarity of the three, we will cover only the HDLC protocol in this discussion. But first, we need to discuss the BISYNC protocol.

The BISYNC Protocol
BISYNC stands for BInary SYNChronous communications; it is a half duplex character-oriented protocol. The protocol has a very rigid format that uses special characters (ASCII or EBCDIC) to delineate the various fields of a message and to control the required protocol functions.

A typical BISYNC message, shown in Figure 9-21, consists of the following discrete parts: (1) two or more synchronizing characters (SYNC), (2) start

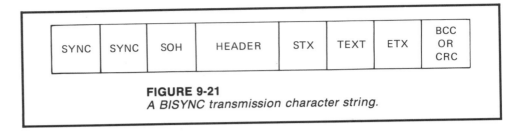

FIGURE 9-21
A BISYNC transmission character string.

of header (SOH), (3) header, (4) start of text (STX), (5) text, (6) an end of text (ETX), and (7) a trace or tail block (BCC or CRC).

The synchronizing (SYNC) characters are used to establish the correct timing between the transmitter and receiver. The number of SYNC characters varies with the different communication applications and networks. The start of header is a format control character that is transmitted just before the header character to identify the individual message control characters. The header is an optional character that normally contains routing or message priority information. The start of text (STX) is a special format control character that is transmitted before the first data characters; it indicates that the characters to follow are information. The text, of course, is the data that is being transmitted. The end of transmission block (ETB) is a format control character indicating the end of text and the beginning of the trace or tail (CRC). It normally used to indicate end of an intermediate text block.

The end of text (ETX) is also a special format control character that indicates the end of a text block and the start of the trace or tail (CRC). The trace or tail block (CRC) detects and corrects errors in transmission. It depends on the information code being used, such as ASCII or EBCDIC, and has a block check character or a combination of checks.

If the trace or tail block is in ASCII, a VRC/LRC message check is performed. In EBCDIC transmission, normally no VRC/LRC check is performed, and the CRC is calculated on the entire message.

The BISYNC is a rather simple protocol, but several problems complicate it. The format of the protocol places special meaning on the ETX character. If the data block (or control information) contains this character among its data, the characters can be misinterpreted. For example, if the datum happens to be an 8-bit pattern identical to the ASCII pattern for ETX, this datum character could deceive the receiver into taking an end-of-block action when actually the message block has more characters to follow. The correct this problem, the protocol needs to distinguish specific patterns as data characters. The ability to treat control characters either as control information or as data is called data transparency.

BISYNC uses the control character DLE (data link escape) to obtain the required transparency. If a control symbol is to be treated as data, it is preceded by the DLE character. In other words, the receiver is warned by the receipt of a DLE to accept the next character as data and not to take any control action. The use of the DLE character is somewhat more complicated than described because of other special considerations. For example, to maintain transmission synchronization in the absence of data in the transmitter queue, the protocol provides for an automatic insertion of sync characters, which are ignored by the receiver. But it is possible that a sync character might be inserted between a DLE and a control character that follows it. This condition forces the receiver to interpret the sync character as if it were data, rather than accepting the next character as data. In this situation, the transmitter cannot simply send a sync character after a DLE. The transmitter must queue the DLE, and then send sync characters until both the DLE and its corresponding data character are ready. If message buffering is not available, the transmitter has to send a DLE and then stay idle on the line. In the situation, it should transmit in the idle state the characters DLE and SYNC, which the receiver can ignore as pairs of characters. Also note that since the DLE is a control character with special significance, the DLE character in the data transmission must be treated the same way the ETX is treated, and it must be preceded by a DLE character when transmitted over a communications link.

An example problem will help to illustrate the BISYNC protocol.

EXAMPLE 9-3

Problem: Assemble the bit streams required to send the message M1 ON using 8-bit odd parity ASCII code and BISYNC protocol. Assume the header is given by 00000001 for transmitting station 1 and omit the check character (BCC or CRC) at the end of the transmission.

Solution: Use the ASCII code table to find the bits required as follows: SYNC = 00010110, SOH = 00000001, STX = 00000010, M = 11001101, 1 = 00110001, space = 00100000, 0 = 01001111, N = 11001110, ETX = 10000011.

SYNC	SYNC	SOH	HEADER	STX	M
00010110	00010110	00000001	00000001	00000010	11001101

1	SPACE	0	N	ETX
00110001	00100000	00000001	00000001	10000011

The HDLC Protocol   The main feature of HDLC protocol is that it opens and closes each message block or frame with start-frame and stop-frame characters (flags). A typical HDLC message is shown in Figure 9-22 and consists of six discrete parts as follows: (1) open flag, (2) address byte, (3) control byte, (4) data field, (5) a check field, and (6) a close flag. The open flag always consists of the same 8 bits (01111110); it is used to indicate the start of a transmission

OPEN FLAG 01111110	ADDRESS	CONTROL	DATA	CHECK FIELD	CLOSE FLAG 01111110

FIGURE 9-22
An HDLC transmission string.

frame. This 8-bit sequence is never repeated again throughout the entire message until the close flag.

The address byte is the address of the transmitter on a command message or the receiver on a response message. The address byte consists of 8 bits allowing for 256 addresses, but 16 addresses is the maximum normally used since some bits are used for other functions.

The control byte has 8 bits and contains command or control response information. The data field, on the other hand, can contain any number of bytes and is the user's total data transmission. The data field normally uses EBCDIC, ASCII, BCD, or straight binary code.

The check field follows the data and precedes the closing flag. It contains a cyclical redundancy check (CRC) character that detects and in some cases corrects errors during transmission. To review, CRC functions as follows: the transmitter sends its computation to the receiver, which compares the transmitted computation with its own calculation. If equal, the data is assumed to be error-free, and, if unequal, the receiver may not accept the transmission and normally requests retransmission.

The close flag is the final transmission byte and has the same bit configuration as the open flag. It terminates the transmission frame and begins the next if its available.

The synchronous nature of this protocol forces the transmitter to have data ready in a buffer at the beginning of a transmission block. If it is not ready and the system fails to produce the data in time for transmission, the transmitter will run out of information to transmit. The HDLC protocol does not have an idle character within a block, so the system must abort an entire transmission block when the transmitter runs out of data to send. The abort code is normally a sequence of eight 1s.

Local Area Networks   A local area network (LAN) is a user-owned and operated data transmission system operating within a building or set of buildings. LANs allow a great number and variety of machines and processes to exchange large amounts of information at high speed over a limited distance. LANs connect communicating devices, such as computers, programmable controllers, process controllers, terminals, printers, and mass storage units within a single process or manufacturing building or plant.

LANs allow neighboring computers to share data resources, hardware resources, and software. For example, in a typical manufacturing facility the process control system computers and a central computer used by purchasing might be tied together to speed up the ordering of raw material for the plant.

A typical LAN for an industrial facility is shown in Figure 9-23.

LAN Topologies   There are three LAN topologies in common use: (1) ring, (2) bus, and (3) star. The *ring* network topology shown in Figure 9-24 can operate using simplex transmission medium. However, most ring topology networks use bi-directional transmission, allowing messages to flow in the most efficient manner. Each node decides whether to accept or pass on a message. This scheme is relatively easy to implement.

The *bus* network topology shown in Figure 9-25 requires a broadcast medium where signals flow to all stations at all times. All the stations receive transmissions, even if they act only on some. The advantage of this scheme is that the stations associated with the bus do no message routing because the bus is a broadcast medium.

The *star* network shown in Figure 9-26 allows only one station to be in communication with the central station at one time. The central station may be allowed to transmit to several nodes at the same time. Routing messages is very easy because the central station has a unique hardware path to each node. System security is high, since access to the network is controlled by the central station. Another advantage is that priority status can be assigned to selected nodes in the network.

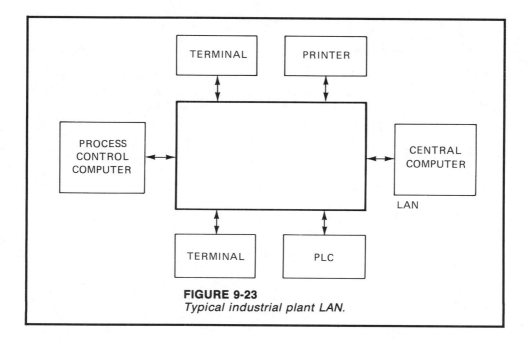

FIGURE 9-23
Typical industrial plant LAN.

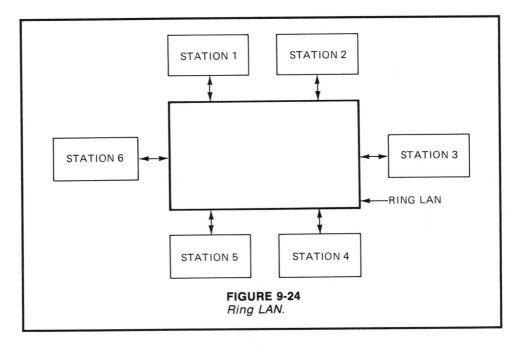

FIGURE 9-24
Ring LAN.

LAN Protocols   A communication protocol was defined earlier as the set of
rules that govern data communications. One of the most important functions
of a LAN protocol is to govern access to the communications network. Failure
to control access would result in chaos any time the traffic on the network rose
above a certain minimum level.

The standard LAN protocols are: (1) polling, (2) token passing, and (3)
carrier sense multiple access/collision detect (CSMA/CD). In *polling,* a master
station selects each of the other stations in turn and gives the station per-
mission to communicate for a fixed period of time. The main disadvantage of
polling is that each station must remain idle except when selected by the
master. This type of protocol is best suited for bus- or star-type LAN topologies,
but it is not normally used because stations must remain idle for a large per-
centage of the time.

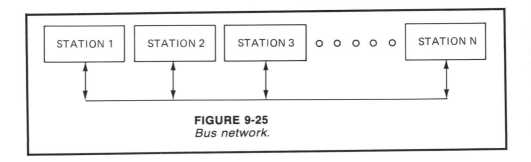

FIGURE 9-25
Bus network.

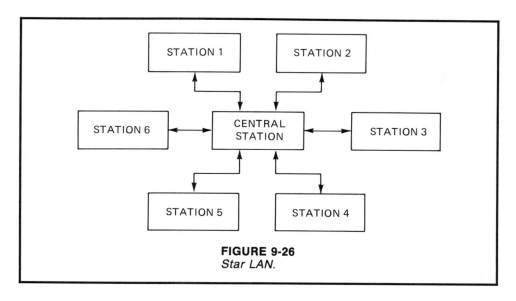

FIGURE 9-26
Star LAN.

The token passing protocol operates by passing a symbolic electronic token from one station to another in the network. Each station may hold the token for a predetermined length of time before passing it. The station that has the token controls the right to transmit to the network. This type of protocol is best suited to a ring-type LAN topology.

The carrier sense multiple access/collision detect (CSMA/CD) protocol allows stations to try to communicate whenever they need access. When a station has a message to send, it first listens to determine if anyone else is transmitting ("carrier sense"). When the station defects an idle channel, it transmits data. If two stations detect an idle channel and transmit simultaneously, a collision will occur. In this case, the "collision detect" part of the protocol informs both stations that the communication has failed; then both stations wait a random amount of time before attempting to retransmit. This type of protocol is generally used in the ring topology.

Standard Network Architecture
Data communications between computer systems and networks is possible only if they adhere to some common set of rules for both hardware and software. A standard approach to network design or architecture that defines the relations between network services and functions is required in computer system design.

The International Standards Organization (ISO) recognized the need for standards to govern the information exchange between and within networks and across geographical boundaries. The standard, which has gained wide acceptance, is a seven-layer model for network architecture, known as the ISO Model for Open System Interconnection (OSI).

The layered approach to network design comes from the design of operating systems. Due to the complexity of most computer operating systems, they are generally designed in sections, each of which contributes a certain function to the operating system. This method of design makes it easier for each section to be refined and redesigned to meet its functional purpose. Finally, all the layers are integrated to provide a totally functioning operating system.

The same procedure can be used to design a network system. The ISO/OSI model specifies a hierarchy of independent layers that contain modules for performing defined functions. The ISO/OSI model has seven distinct layers, at both the receiver and the transmitter, through which communications must pass, as shown in Figure 9-27.

The function of each layer of the ISO/OSI model is described as follows:

1. The *physical* layer defines the electrical and mechanical requirements of interfacing to a physical medium for transmitting information. When used, this layer must include the software driver for each communications device and the hardware such as interface equipment, modems, and the communication cables.

2. The *data link* is used to establish an error-free communications link between network stations over the physical channel. It formats messages for

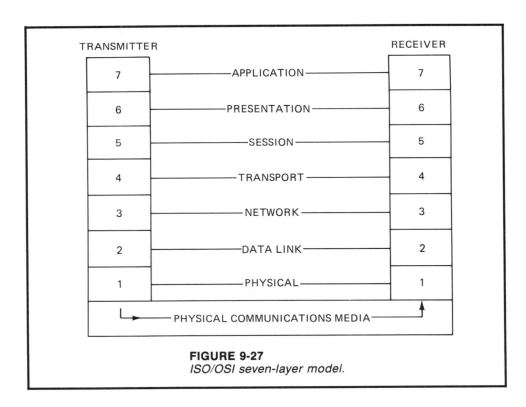

FIGURE 9-27
ISO/OSI seven-layer model.

transmission, checks integrity of received data, controls access to and use of the station, and ensures the proper sequence of the transmitted information.

3. The *network control* layer is used to address messages, set up the path between stations, route messages across intervening stations to their destinations, and control the flow of data between stations.

4. The *transport* layer furnishes end-to-end control of a communication once the path has been established, allowing the system to exchange data reliably and sequentially. This layer is normally beyond user control.

5. The *session* layer organizes the dialog of the communication and manages data exchange. This layer is also normally beyond user control, as are layers 6 and 7.

6. The *presentation* layer handles tasks related to data representation and code conversion.

7. The *application* layer handles tasks related to data transfer speed and integrity.

Outside the control of the ISO/OSI are the communications applications processes. Usually, an application process is at the transmitter and another is at the receiver. Computer operations that require a user at a terminal or a piece of software performing instructions are examples of application processes.

EXERCISES

9.1 Discuss in detail the function of the three basic components of a communications system.

9.2 Explain the operation of simplex, half duplex, and full duplex communications and give the advantages and disadvantages of each method.

9.3 Compare and discuss serial and parallel data transmission. Give the advantages of each method and some typical applications encountered in computer systems.

9.4 Discuss the three common methods of signal multiplexing encountered in communications systems.

9.5 Generate the ASCII code with odd parity for the message LEVEL HIGH ON TANK 200.

9.6 Explain the four common data transmission error checking methods used in programmable controller communications systems.

9.7 Calculate the LRC/VRC odd parity characters required for the message: MOTOR 1 OFF.

9.8 Explain the difference between asynchronous and synchronous data transmission and given an example of each method.

9.9 Discuss the three common serial data protocols used in communications systems.

9.10. Assemble the bit streams required to send the message MOTOR M-125 ON using 8-bit even parity ASCII code and BISYNC protocol. Assume the header is given by 00000010, which represents station 2, and use a CRC check character at the end of the transmission.

9.11 Discuss the function of each of the seven layers of the ISO/OSI communications standard.

BIBLIOGRAPHY

1. *Reference Manual Data Highway/Data Highway Plus Protocol and Command Set*, Allen-Bradley Company, Inc., 1987.

2. *Introduction to Local Area Networks*, Digital Equipment Corporation, 1982.

3. Seyer, M. D., *RS-232 Made Easy: Connecting Computers, Printers, Terminals, and Modems*, Prentice-Hall, Inc., 1984.

4. Stone, H. S., *Microcomputer Interfacing*, Addison-Wesley Publishing Company, 1982.

5. Thiel, C. A., (Ed.), *IBM Systems Journal Telecommunications*, Volume Eighteen, Number Two, International Business Machine Corporation, 1979.

6. *Digital Industrial Networks Guidebook*, Digital Equipment Corporation, 1988.

7. *Networking: The Competitive Edge*, Digital Equipment Corporation, 1985.

8. Stacy, A. H., *The Map Book: An Introduction to Industrial Networking*, Industrial Networking Incorporated, 1987.

10

Applications

Introduction   As a general type of control system, the programmable controller offers a wide variety of system configurations and capabilities. These range from a single machine relay replacement to an entire process control and monitoring system. After the decision has been made to use a programmable controller in an control application, the design engineer must complete the system design.

System Design   The design of a programmable controller-based control system requires a simplified process or mechanical flow diagram, a process control description, sizing and selection of a PC, a system specification, system drawings, wiring diagrams, and control programming. Each of these areas will be discussed to give a step-by-step approach to system design.

Process and Instrument Drawings   The design of any control system must start with the process and instrument drawing (P&ID) or mechanical flow diagrams (MFDs). These drawings show the process and/or mechanical equipment to be controlled and the instrumentation used in the control of the process or machine.

In the process industry, a standard set of symbols is used to prepare the P&IDs and MFDs. The symbols used in these drawings are generally based on Instrument Society of America (ISA) standard, ANSI/ISA-S5.1-1984, Instrumentation Symbols and Identification. These drawings show the interconnection of equipment and the instrumentation used to control the process. A typical example of a P&ID is shown in Figure 10-1.

In standard P&IDs, the process flow lines, such as process fluid flow and steam flow, are shown as heavy solid lines. The instrumentation signal lines are shown in a way that indicates whether they are pneumatic or electric. A

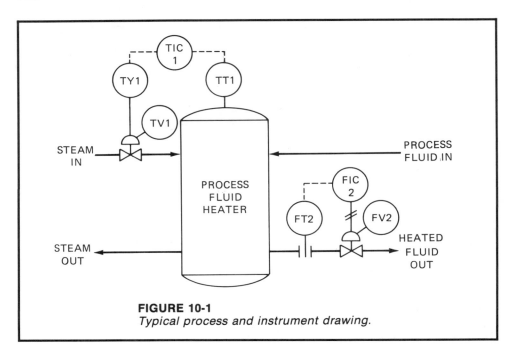

FIGURE 10-1
Typical process and instrument drawing.

cross-hatched line is used for pneumatic lines, for example, a 3-15 psi signal. The electric current line, usually 4-20 mA is represented by a dashed line.

A balloon symbol with an enclosed two- or three-letter code is used to represent the instrumentation associated with the process control loops. For example, the balloon in Figure 10-1 with TT1 enclosed is a temperature transmitter, and that with TIC enclosed is a temperature-indicating controller. Generally, a number is assigned to each control loop; combining the letter code and number into an instrument tag number labels the specific device in the loop, as illustrated in Figure 10-1.

Special items such as control valves and in-line instruments (for instance, orifice plates) have special symbols, as shown in Figure 10-1. Refer to ISA-S5.1 for a more detailed discussion of instrument symbols.

The control system design engineer normally reduces the P&ID to a simplified process control diagram that shows only the equipment and instrumentation controlled or measured by the programmable controller. These simplified diagrams are then used to show the status of the process in each step or state to aid in the programming of the system.

Process Control System Description   The process description is probably the most important step in the design process because it conveys in simple language the purpose and the steps of the process. It is important because in most applications it is the main vehicle of communication between the user and the designer.

Sizing and Selecting a PC System The sizing of a PC system consists of estimating the number of input/output (I/O) modules required to control the process and the size of PC memory needed. The selection process consists of choosing the correct programming language and peripheral equipment required for the control application.

I/O Sizing Many different types of I/O modules may be needed for a given PC application. Limit switches, push buttons, selector switches, motor controls, solenoids, and pilot lights may require either ac or dc modules of different voltage levels. Solid-state displays and some electronic instrumentation may require $+5$ V dc logic interface modules. Process instrumentation for measurement of level, flow, temperature, or pressure may require analog-to-digital (A/D) conversion interfaces. Incremental encoders and stepping motors might need pulse-type input/output modules.

The interface to these I/O devices can be done with external "black boxes" to condition the signals, which would increase overall equipment cost for a system. Therefore, a PC should be selected with the correct input/output modules to match the process equipment.

The number of field devices that can be interfaced to a PC system is an important consideration in sizing the I/O. Each programmable controller has a maximum number of input and output devices that can be monitored or controlled. Most PC systems have I/O capacities ranging from 16 to 4096. These capacities can be divided into three categories: small, medium, and large. A small PC system would range from 10 to 64 I/O, a medium PC system encompasses 64 to 1024 I/O, and a large PC system has 1024 to 4096 I/O. To determine the PC system size required, simply add up the number of field and control panel devices and compare the total to the above classifications. The system designer must also define the type and number of I/O, because some PC systems are constrained as to the mixture that can be interfaced to a given I/O system.

Input and output totals can then be used to determine the number and type of I/O modules required. Each module can interface a certain number of I/O, such as 2, 4, 8, or 16. Divide the number of inputs or outputs by the number of I/O points per module and round up to the nearest whole number. This calculation must be performed for each type of I/O module, i.e., 120 V ac, 24 V dc, pulse, analog, etc.

After defining the I/O requirements, the designer must also consider spares and future expansions. Most programmable controller users find that 10% to 20% spare capacity is enough for normal system growth.

Memory Sizing The amount of memory required for an application is primarily a function of control program complexity and the number of I/O points in the system. The most precise method of determining memory size is to write out the control program and count the number of instructions used in the program. Then multiply this count by the number of words used per instruction. (This number can be obtained from the PC programming manual. Also consult the programming manual on the amount of memory used by executive programs and the processor scratch pad.) The problem with this method is that, in a large system, the system programming sometimes takes months to complete, and the system must be designed and purchased in advance. (A shortcut method was given earlier that consists of multiplying the number of I/O points by 10 to obtain the memory required.)

An illustration of memory sizing using this simpler method is given in Example 10-1.

EXAMPLE 10-1

Problem: Calculate the memory size required for the PC application shown in Figure 10-2.

Solution: To calculate the memory size, we first need to calculate the number of I/O points in the system in the figure.

Remote Area 1 I/O points = 70 + 35 + 6 = 111
Remote Area 2 I/O points = 95 + 50 + 10 = 155
Main Process I/O points = 300 + 156 + 32 + 5 = 493
Total system I/O points = 111 + 155 + 493 = 759
Therefore, memory size = 10 × 759 = 7590 ≅ 8K

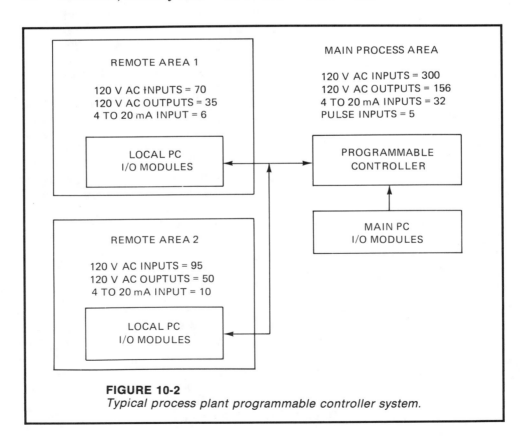

FIGURE 10-2
Typical process plant programmable controller system.

Selecting Programming Language   As mentioned earlier, the four basic types of programming languages available in programmable controller systems are ladder diagrams, Boolean, function blocks, and sequential function charts. The type selected depends on the complexity of the control system and the background of the control system programmer and operators. Most PCs offer the basic relay ladder logic instructions plus a combination of the other types of languages. The most common type of language selected is a combination of ladder logic and function blocks, since this covers the basic ladder logic and some data transfer and manipulations.

Peripheral Requirements   The term "peripheral" refers to the equipment in the programmable controller system that is not directly involved in process control but increases the capabilities of the system. The most common peripheral is the programming device. This is generally available in three formats: a compact portable panel, a CRT (cathode ray tube) with a keyboard, or an industrial-type personal computer. The compact programmer normally has a small, limited function key pad and a seven-segment LED display and can handle only one logic rung at a time. It is normally used on small systems or for minor field changes to larger systems. The CRT-type programming device is relatively portable and will normally display several rungs or logic functions at one time. The personal computer-based programmer has the greatest number of programming functions, but it is normally not portable and is generally used in a lab or office environment to perform the programming on PC systems.

Another common peripheral used in PC systems is a magnetic tape cassette unit. This unit is used to store the control program on magnetic tape so that, in the event of a program loss, the backup program can be reloaded into the controlled memory. If a personal computer is used in the PC system, the program can be saved on a floppy disk or a hard disk for future use.

For hard copy printouts of the control programs, a printer can be interfaced with the programming device to obtain a program listing. Some software programs can document the control program on special industrial computers or personal computers that interface to a programmable controller system.

It is important for a complete system design that the peripheral equipment be available to back up and document a control program, because manual reprogramming and documentation can be expensive and time consuming.

Other common peripherals are PROM programmers, process I/O simulators, and communications modules. The PROM programmers are used to write and save control programs on the PROM chips used in some controllers. The I/O process simulators are useful and time-saving devices for large and complex systems that can be fully tested before installation and start-up in the field. The communication peripherals are used to communicate between the programmable controllers and the plant or personal computers and other controllers in a system.

System Drawing and I/O Wiring Diagrams   A system drawing is used to given an overall view of the system hardware (I/O modules, processor, and peripheral equipment) and the system interface cabling. This drawing is also useful in identifying all the interface cables by model number. A typical system drawing is shown in Figure 10-3. The system consists of a programming controller, a programming terminal, a printer, two I/O communications cables, two dc power supplies, two I/O equipment racks, and the associated input/ output modules.

A typical I/O wiring diagram is given in Figure 10-4. This drawing shows the wiring of the process equipment, such as heaters, pumps, and motors, to an ac output module. The wiring terminal strip in the programmable controller equipment cabinet is designated by TB-1, and a field junction box terminal strip is designated by JB-1 in the example. Field junction boxes are used extensively in process control applications to simplify field installation and maintenance of instrumentation. Field wiring is normally indicated on wiring diagrams by a dashed line, as shown in Figure 10-4. The PC output addresses are given on the left-hand side of the wiring diagram.

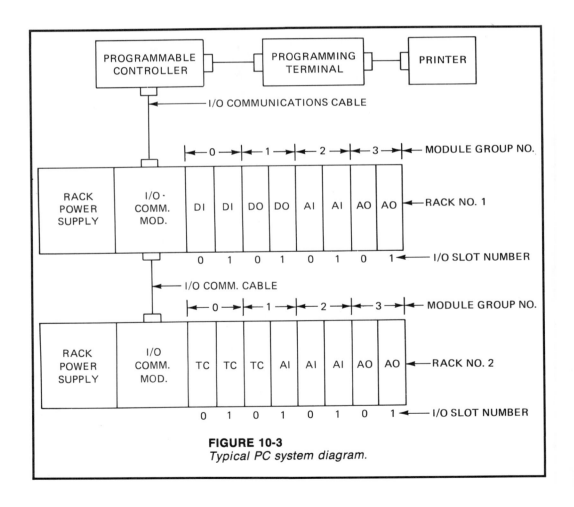

FIGURE 10-3
Typical PC system diagram.

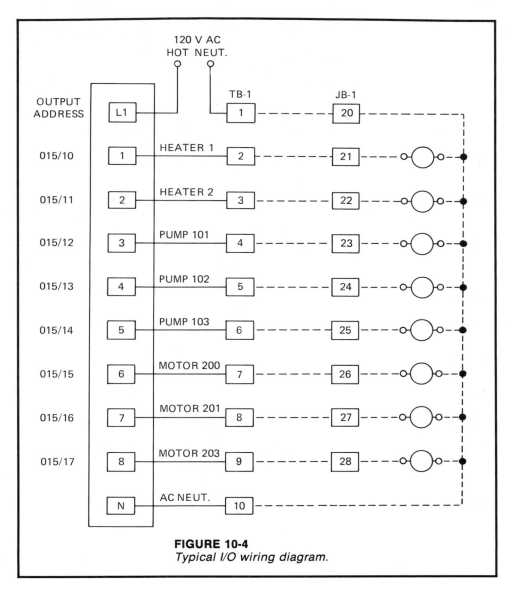

FIGURE 10-4
Typical I/O wiring diagram.

It is recommended that the I/O program address number be put on the wiring diagrams for easier troubleshooting of the control program and the field wiring during installation and start-up.

System Programming   The programming of a PC system can be done by the design engineer, process operations (plant) personnel, or the control system vendor. Programming by the system design engineer is the best choice, since it requires less documentation (flowcharting, process description, etc.) and less time. The second best choice is to have the plant or operating personnel perform the system programming. More up-front documentation and time are required, but this method results in greater control system acceptance by plant personnel. Programming by a systems vendor is the poorest choice, since it requires more documentation and debugging time and more training time for plant personnel.

The selection of programming language type should usually be left to plant operations personnel, since they will normally have to maintain the software after the system is installed.

Application   We are now ready to discuss a PC-based control system application. The design of a PC-based control system requires a process flow diagram or a P&ID, a process control description, sizing and selection of a PC, a system specification, system drawings and wiring diagrams, and control programming.

These design steps can be illustrated by using a PC to control the dehydration process shown in Figure 10-5. The dehydration process removes water from natural gas by using small beads in the process tower to absorb moisture from the gas.

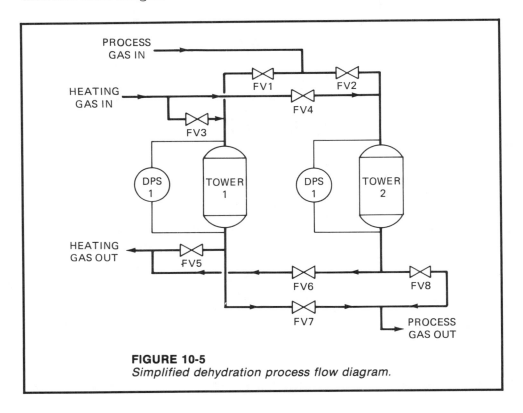

FIGURE 10-5
Simplified dehydration process flow diagram.

In this process, air-operated valves (AOVs) are used to control the gas flow to the two towers shown. A typical diagram of the AOVs is shown in Figure 10-6. An electrically operated solenoid valve is used to supply plant air at 80 psi to turn on the control valves, and the valves have limit switches to indicate if the valve is open or closed.

The limit switches are needed by the programmable controller to determine the status or state of the process. Normally, only one limit switch would be required, since the programmable controller can assume the opposite state (open or closed) if only one switch is used. However, a more reliable control is obtained if two switches are used. For example, if a valve fails to completely open or close on a given operation, the programmable controller is able to detect this failure and signal the operator.

We now have a simplified process flow diagram we can use to show the valve and/or equipment position for each step or state of the process. The next step is to write a preliminary process control description.

Process Control Description   The two process towers (towers 1 and 2) are used to remove moisture from natural gas. Generally, one tower is *in service* (i.e., removing moisture from the process gas) and the other tower in being dried out or *regenerated*.

The steps of the process are as follows:

1. The operator selects the tower to be placed in service first by using hand switch 3 (HS/L-3) or hand switch 4 (HS/L-4) on the control panel shown in Figure 10-7.

2. Then the operator depresses the system start button (HS/L-1)

3. If we assume that tower 1 was placed in service first, process gas is fed to tower 1 by opening flow control valves FV-1 and FV-7, and keeping valves FV-2, FV-3, FV-4, FV-5, FV-6, and FV-8 closed.

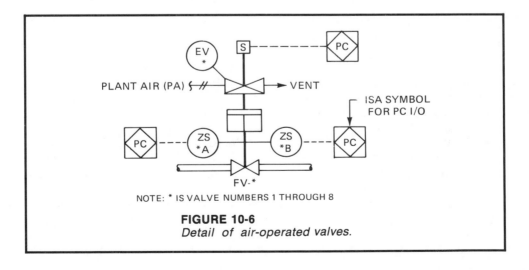

NOTE: * IS VALVE NUMBERS 1 THROUGH 8

FIGURE 10-6
Detail of air-operated valves.

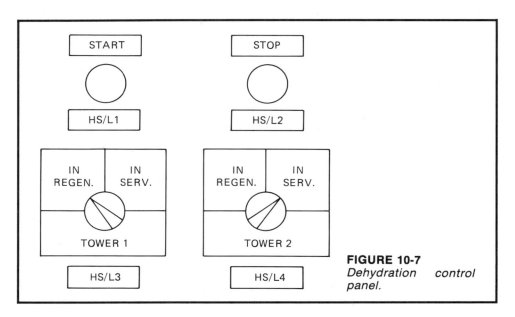

FIGURE 10-7
Dehydration control panel.

4. When the differential pressure (dP) across tower 1 becomes high, as indicated by differential pressure high switch (dPHS-1), the programmable controller places tower 1 in the regeneration mode by opening valves FV-3 and FV-5 and closing FV-1 and FV-7. At the same time, tower 2 is placed in service by opening valves FV-2 and FV-8. Note at this time that the heating cycle valves for tower 2, FV-4 and FV-6, are still closed.

5. Later, if the differential pressure across tower 2 becomes high and dP signal on tower 1 returns to low, the control system will place tower 1 in service and tower 2 in regeneration by opening valves FV-1, FV-7, FV-4, and FV-6 and closing valves FV-3, FV-5, FV-2, and FV-8.

After this preliminary control description is written, the designer is ready to size and select a programmable controller system.

Sizing and Selecting the Application System   The number of inputs and outputs can be estimated from the dehydration flow diagram in Figure 10-5 and the dehydration control panel in Figure 10-7. There are eight (8) 120-V ac solenoid valves (eight 120-V ac outputs) and sixteen (16) limit switches (sixteen 120-V ac inputs) for the open and closed positions of the valves. There are also four differential pressure alarm switches and four control panel switches for eight (8) additional 120-V ac inputs. We arbitrarily selected 120-V ac for the input signal voltage to save on the types of modules and supply voltages used in the system. Hand switches HS1, HS2, HS3, and HS4 also have six ac indicator lights (i.e., start, stop, tower 1 in regeneration, tower 1 in service, tower 2 in regeneration, and tower 2 in service), so there are six more 120-V ac outputs. This results in twenty-four 120-V ac inputs and fourteen

120-V ac outputs for a total I/O count of 38. Based on our size classifications for PC systems, this application is classified as a small system.

Let us assume that we select 8-point, 120-V ac I/O modules for use in our application. The number of modules required can be calculated as follows:

- 120-V ac input modules
  $(24 + 20\%) \div 8$ points/module $\cong 4$ modules

- 120-V ac output modules
  $(14 + 20\%) \div 8$ points/module $\cong 2$ modules

System Drawing   The system drawing for this application will consist of a programmable controller, a programming terminal, a printer to document the program, and a single I/O rack with 8 slots for I/O modules, as shown in Figure 10-8.

The system diagram is normally drafted on a "D" size (24-in. by 36-in.) drawing sheet and will include a detailed material list. This list includes the equipment number, material description, manufacturer, and model number, as shown in Table 10-1.

The placement of the ac input (ACI) modules and ac output (ACO) modules in the I/O equipment rack is quite arbitrary in this application since we have selected only ac modules. If the system has a mix of ac and analog modules, the ac modules are normally integrated from low signal dc modules such as 4 to 20 mA dc, 5 V dc, and mV inputs to avoid electrical interference.

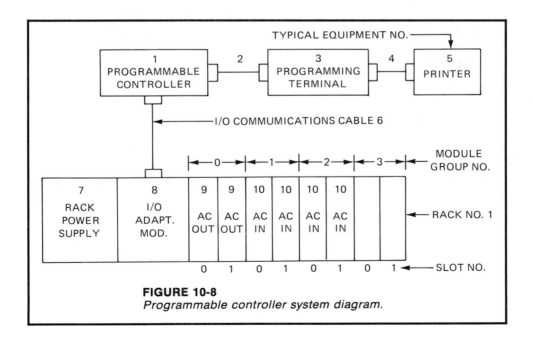

FIGURE 10-8
Programmable controller system diagram.

TABLE 10-1
Dehydration Control System Material List

Eq. No.	Material Description	Manufacturer	Model No.
1	Programmable controller	PC Systems	PC1000
2	PC to terminal cable	"	C-400-1
3	Programming terminal	"	T-300A
4	Printer cable	"	CC-80X1
5	Graphics printer	"	GP-80MX
6	I/O Communications cable	"	CC-1000
7	Rack power supply	"	PC-5A
8	I/O adapter module	"	ADP
9	120-ac output module	"	ACO-120
10	120-ac input module	"	ACI-120

I/O Wiring Diagrams The I/O wiring diagram for the first ac output module is shown in Figure 10-9. This drawing shows the wiring of the system solenoid valves to the ac output module.

The wiring diagram for the ac output module in module group 0, slot 1 is used to drive the lights on the control panel (see Figure 10-10).

In our example application, a single ac line is used where the ac hot is normally designated by L1 and ac neutral has wire number L2. However, in larger systems, more than one ac line might be used, depending on loading requirements. The design engineer must also consider maintenance of the system, so that in our example application we might connect the wiring for tower 1 to one ac circuit and the wiring to the other tower to a second ac circuit. In this case, one tower could be taken out of service and the other tower could be kept on line during maintenance. Furthermore, the wiring for the control panel might be placed on a third ac power circuit.

The wiring of the ac input modules is performed in a similar manner, except that the field inputs are normally drawn on the left side of the input module as shown in Figure 10-11. This input wiring diagram shows the connection of the differential pressure high- and low-level switches. As shown in the system drawing, the first ac input module is in module group 1, slot 0, so that the input addresses are between bit 111/00 and bit 111/07, as shown on the right-hand side of the module input drawing.

The input wiring diagram in Figure 10-12 shows the connection of the valve limit switches for control valves FV-1 through FV-4. As shown in the system drawing, this ac input module is in module group 1, slot 1, so that the input addresses are between bit 111/10 and bit 111/17, as shown on the right-hand side of the module input drawing.

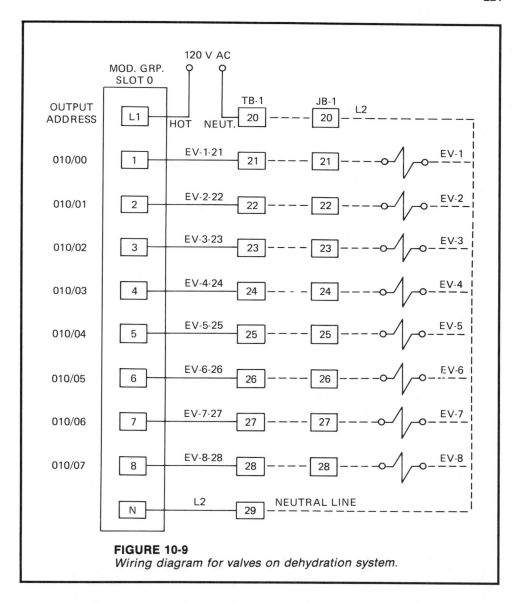

FIGURE 10-9
Wiring diagram for valves on dehydration system.

The wiring diagram for the limit switches on the remaining four control valves (i.e., FV-5, FV-6, FV-7, and FV-8) is shown in Figure 10-13.

The final wiring diagram needed is for the panel-mounted push buttons and hand switches (see Figure 10-14).

Application Programming   In a small system like our dehydration application, basic ladder programming is the best choice, since the system requires

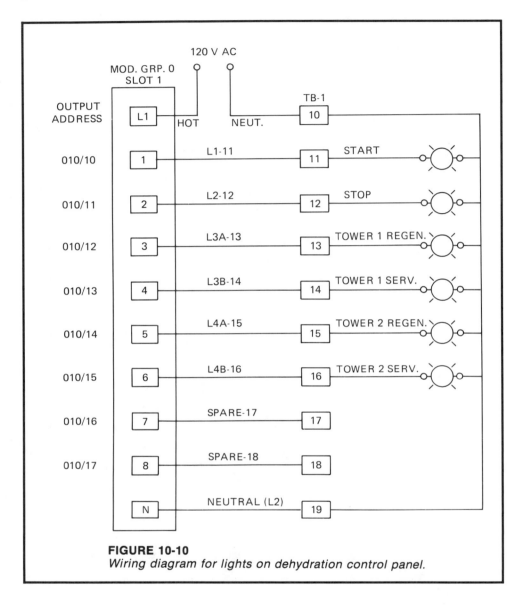

FIGURE 10-10
Wiring diagram for lights on dehydration control panel.

only simple on/off control and there are no involved analog or data manipulations anticipated.

The first step in the logic programming is to make a list of the I/O points in the system using the I/O module wiring diagrams. The I/O data table bit assignments for our application are given in Table 10-2. The I/O bit assignment for the inputs on the system are given in Table 10-3.

These bit assignment tables are an aid in programming because they consolidate the I/O information in a single table for easy reference.

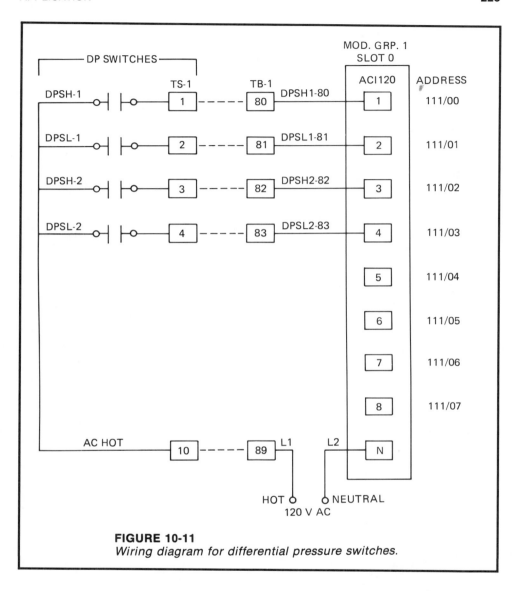

FIGURE 10-11
Wiring diagram for differential pressure switches.

The first step in writing the control software is to design the ladder logic needed to start the process. Figure 10-15 shows the logic required to turn on the panel-mounted start light for the system.

The operation of this ladder program is relatively simple. When the start push button is depressed on the control panel, input bit 112/00 is true and bit 112/11 is true since the STOP push button is normally closed. So there is logic continuity in ladder rung one (1). This activates output coil 0010/00, which seals the ciruit ON after the start push button is released and turns on the start light on the control panel. The start coil can be deactivated by depressing

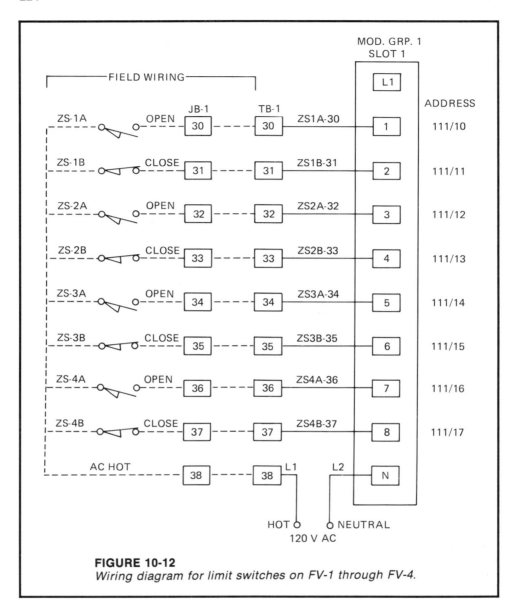

FIGURE 10-12
Wiring diagram for limit switches on FV-1 through FV-4.

the stop push button. The normally closed contacts from the start coil instruction are used to activate the STOP light at bit location 010/11 in rung 3 of the program.

The next step in the software design is to write the logic for the control of the on/off valves in the process. Each tower in the system can be only in

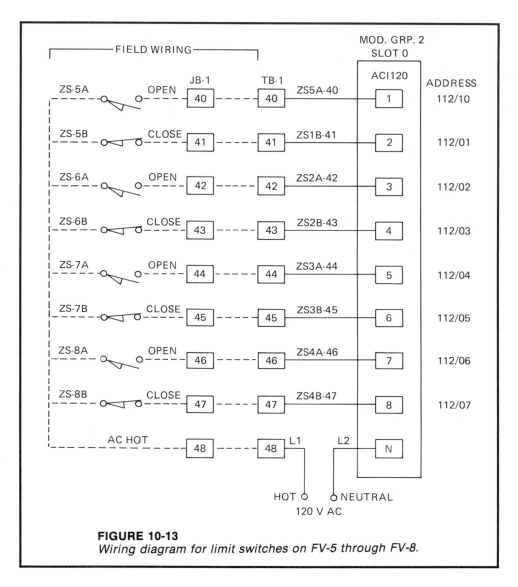

FIGURE 10-13
Wiring diagram for limit switches on FV-5 through FV-8.

one of two states, either in service (In Serv.) or in regeneration (In Regen.). So we can write the ladder logic for Tower 1 in service, as shown in Figure 10-16. In this example, we set the output bit 010/12, which turns on the "Tower 1 In Service" light on the control panel.

As shown in Figure 10-16, Tower 1 is placed in service if the differential pressure in the tower is low, the system start coil (bit 010/10) is true, and the tower 1 differential pressure is not high (bit 111/00) and HS-3 selector switch

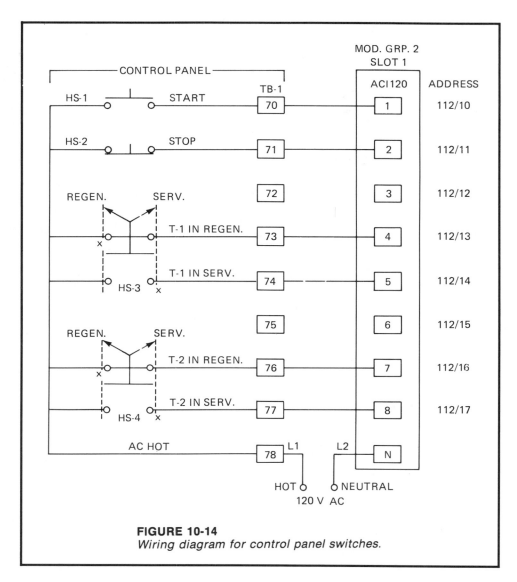

FIGURE 10-14
Wiring diagram for control panel switches.

is in "Service" position. The output for "Tower 1 In Service" is sealed in with its own control bit 010/13. Tower 1 will stay "in service" until the differential pressure in the tower reaches a high level, as measured by a high differential pressure (DPSH-1) across the tower. If the dP switch DPSH-1 is closed, then input bit 111/00 is set to true and its normally closed contacts are opened. So the "Tower 1 In Service" output coil bit 010/13 is turned OFF.

To place Tower 1 into regeneration so that moisture that has built up

TABLE 10-2
Data Table Bit Assignment for Ac Outputs

Word	Bit	Description	Word	Bit	Description
010	00	EV-1 GAS IN T-1		00	
010	01	EV-2 GAS IN T-2		01	
010	02	EV-3 HEAT IN T-1		02	
010	03	EV-4 HEAT IN T-2		03	
010	04	EV-5 HEAT OUT T-1		04	
010	05	EV-6 HEAT OUT T-2		05	
010	06	EV-7 GAS OUT T-1		06	
010	07	EV-8 GAS OUT T-2		07	
010	10	L-1 START		10	
010	11	L-2 STOP		11	
010	12	L-3A T-1 IN REGEN.		12	
010	13	L-3B T-1 IN SERV.		13	
010	14	L-4A T-2 IN REGEN.		14	
010	15	L-4B T-2 IN SERV.		15	
010	16			16	
010	17			17	

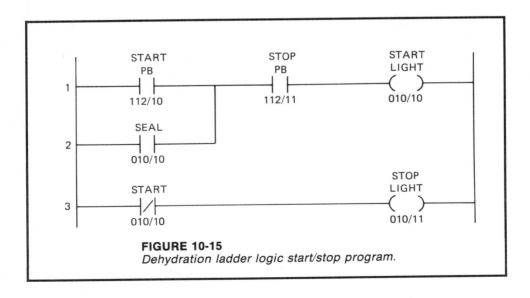

FIGURE 10-15
Dehydration ladder logic start/stop program.

TABLE 10-3
Data Table Bit Assignments for AC Inputs

Word	Bit	Description	Word	Bit	Description
111	00	TOWER 1 dP HI	112	00	FV-5 OPEN
111	01	TOWER 1 dP LO	112	01	FV-5 CLOSED
111	02	TOWER 2 dP HI	112	02	FV-6 OPEN
111	03	TOWER 2 dP LO	112	03	FV-6 CLOSED
111	04		112	04	FV-7 OPEN
111	05		112	05	FV-7 CLOSED
111	06		112	06	FV-8 OPEN
111	07		112	07	FV-8 CLOSED
111	10	FV-1 OPEN	112	10	HS-1 START
111	11	FV-1 CLOSED	112	11	HS-2 STOP
111	12	FV-2 OPEN	112	12	
111	13	FV-2 CLOSED	112	13	T-1 IN REGEN.
111	14	FV-3 OPEN	112	14	T-1 IN SERV.
111	15	FV-3 CLOSED	112	15	
111	16	FV-4 OPEN	112	16	T-2 IN REGEN.
111	17	FV-4 CLOSED	112	17	T-2 IN SERV.

during the service cycle can be removed, we use the logic shown in Figure 10-17.

We are now ready to control the flow valves on Tower 1. The valves are designed to fail closed (FC), so if plant air or electric power is lost, the valve will close. Therefore, the solenoid valves must be energized to open the valve.

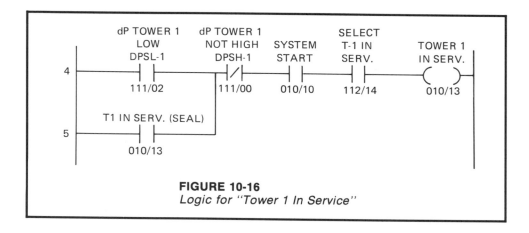

FIGURE 10-16
Logic for "Tower 1 In Service"

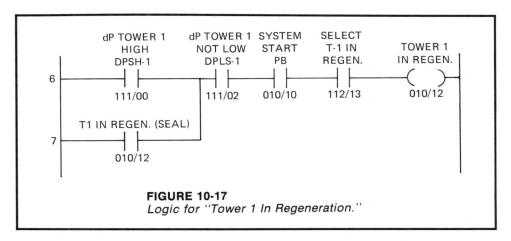

FIGURE 10-17
Logic for "Tower 1 In Regeneration."

The ladder logic program in Figure 10-18 shows the control of Tower 1 electric valves EV-1, EV-7, EV-3, and EV-5.

The same design procedure can be used to write the control software for Tower 2. To this point we have not used the open and closed limit switches in our system program. A typical application for valve limit switches is to animate process graphics displays to warn the operator if the valves do not go to the correct position during any step in the process.

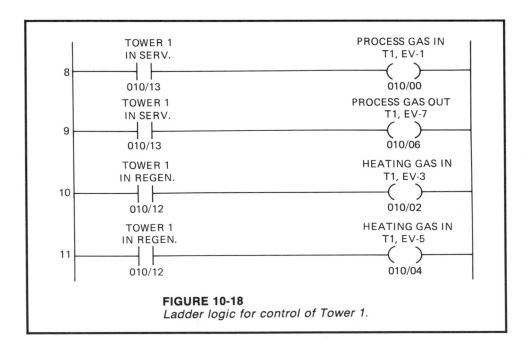

FIGURE 10-18
Ladder logic for control of Tower 1.

Process Graphic Displays

Two methods are commonly used to provide operators with color process graphics displays in programmable controller-based control systems. The first is to hard wire from the programmable controller I/O modules to a graphics display panel with hard wired lights and digital indicators. This method is cost-effective in a small system that will not be changed. It is generally not recommended in larger control systems that will be expanded in the future. The second method is to use an industrial personal computer with process color graphics software. The personal computer-based method has the advantages that the process display screens can be easily modified for process changes, and the computer can perform other functions such as alarm listing, report generation, and programmable controller software.

These features can be best explained by using a personal computer on the dehydration control system. We replace the programming terminal with an industrial grade personal computer and add an RS-232 communications module, as shown in system drawing in Figure 10-19.

The software for vendor-supplied process color displays is normally menu-

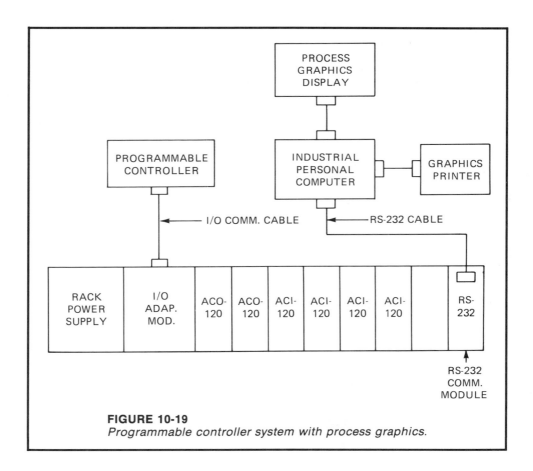

FIGURE 10-19
Programmable controller system with process graphics.

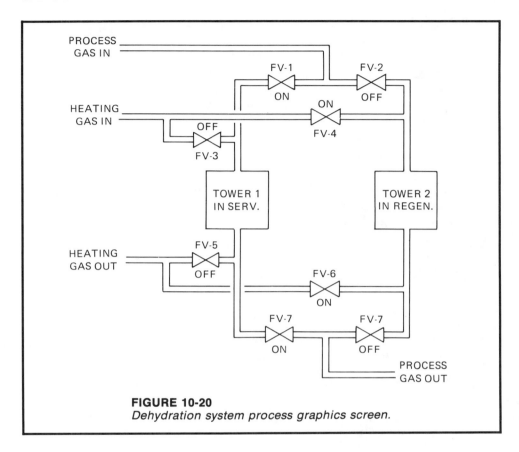

FIGURE 10-20
Dehydration system process graphics screen.

driven and relatively easy to use. The process screens are usually based on the process and instrument drawings for the process being controlled.

For example, if we built a process display based on the dehydration process discussed earlier, the computer-generated display would be as shown in Figure 10-20. One of the differences is that the process lines are normally drawn as double lines so they can be colored in by the graphics program to indicate fluid flow based on signals from the programmable controller. Another difference is that process data and alarm messages can be displayed on the screen.

In the example in Figure 10-20, Tower 1 is shown in service so that valves FV-1 and FV-7 are ON or open and valves FV-3 and FV-5 are OFF or closed. At the same time, Tower 2 is shown in regeneration with the appropriate valves labelled ON and OFF.

EXERCISES

10.1 Write the ladder logic program needed to control Tower 2 of the dehydration system discussed in this chapter.

10.2 Draw a revised process and instrument drawing showing the process equipment and instrumentation needed to provide a third process tower to allow for maintenance of the dehydration process.

10.3 Revise the dehydration control panel to include the hand switches and lights needed to control a third tower on the process. Assume the hand switches will have three positions: In Service, Out of Service, and In Regeneration.

10.4 Write a process description for a revised dehydration system with a third tower added for maintenance.

10.5 Calculate the number of input and output modules required on the dehydration system if a third process tower is added. Assume that all inputs and outputs are 120 V ac.

10.6 Redesign and redraw the I/O module wiring diagrams if a third process tower is added to the dehydration system. Use four separate ac supply lines (L1, L2, L3, and L4), one line for each process tower and a fourth on the control panel.

BIBLIOGRAPHY

1. Jones, C. T., and Bryan, L. A., *Programmable Controllers, Concepts and Applications*, International Programmable Controls, Inc., 1983.

2. Hughes, T. A., *Basics of Measurement and Control*, Instrument Society of America, 1988.

3. *Processor Manual PLC-5 Family Programmable Controllers*, Allen-Bradley Co., Inc., 1987.

4. *Programming and Operations Manual, PLC-2/30 Programmable Controllers*, Allen-Bradley Co., Inc., 1988.

5. *Allen-Bradley Industrial Computer and Communications Group Product Guide*, Allen-Bradley Co., Inc., 1987.

6. *Allen-Bradley Programmable Controller Products*, Allen-Bradley Co., Inc., 1987.

7. Rockis, G., and Mazur, G. *Electrical Motor Controls, Automated Industrial Systems*, American Technical Publishers, Inc., Second Edition, 1987.

11

Installation and Maintenance

Introduction   To complete the discussion of programmable controller system design, we need to cover panel design, equipment installation and layout, system operational testing, and maintenance procedures. The reliability and maintainability of a control system is to a large part a function of proper system design that considers the maintenance aspects of a system.

Control Panel Design   The main feature that separates programmable controllers from other types of computers is that programmable controllers are designed to be installed in harsh industrial environments. They are, in some cases, simply installed on metal sheet (sub panel) on the production floor. However, in most cases, programmable controller system components are installed in a metal enclosure or a control panel to protect against atmospheric contaminants such as dust, moisture, oils, and other corrosive airborne substances. These metal enclosures also reduce the effects of electromagnetic radiation generated by electrical or welding equipment.

The panel or enclosure design requires a panel layout design, considerations for heating and maintenance, wiring layout and duct design, power distribution design, and normally the writing of a panel specification.

Panel Layout   The panel size depends on the amount of equipment to be installed in the enclosure and whether front panel controls and instruments are required on the system. If front panel controls are required, the size and shape of the panel are mainly controlled by the best layout design of these front panel instruments. To assure correct and easy operation of the system, the indicators and recorders are placed at normal eye level and hand switches are placed below the indicators.

The metal enclosure should conform to industrial standards, such as the

National Electrical Manufacturers Association (NEMA) standards. The NEMA standard covers the design of industrial enclosures for different industrial environments per Table 11-1.

The equipment layout inside the panel should follow the recommendations contained in the programmable controller installation manual. In this manual the manufacturer generally lists the minimum spacing allowed between I/O racks, other equipment, and the processor. This spacing generally takes into consideration equipment heating, electrical noise, and safety factors. Figure 11-1 shows a typical minimum equipment spacing for a programmable controller system provided by the PC manufacturer.

Heating Considerations   To allow for effective convection cooling, most manufacturers recommend that all system components be mounted in a po-

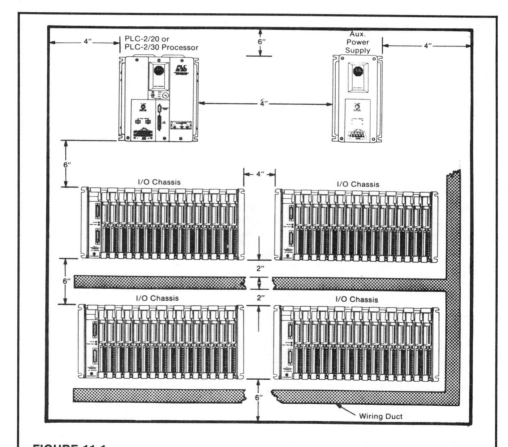

FIGURE 11-1
Typical minimum equipment spacing. (Courtesy of Allen-Bradley, a Rockwell International Co.)

TABLE 11-1
NEMA Type Enclosures for Electrical Equipment (1000 Volts Maximum)

The following descriptions are excerpts from NEMA's
"Standard's Publication/No. 250—1979." [4]

Non-Classified Location Enclosures

Type 1 Enclosures
Type 1 enclosures are intended for indoor use primarily to provide a degree of protection against contact with the enclosed equipment in locations where unusual service conditions do not exist. The enclosures shall meet the rod entry and rust-resistance design tests.

Type 2 Enclosures
Type 2 enclosures are intended for indoor use primarily to provide a degree of protection against limited amounts of falling water and dirt. These enclosures shall meet rod entry, drip, and rust-resistant design tests. They are not intended to provide protection against conditions such as internal condensation or internal icing.

Type 3 Enclosures
Type 3 enclosures are intended for outdoor use primarily to provide a degree of protection against windblown dust, rain, sleet, and external ice formation. They shall meet rain, external icing, dust, and rust-resistance design tests. They are not intended to provide protection against conditions such as internal condensation or internal icing.

Type 3R Enclosures
Type 3R enclosures are intended for outdoor use primarily to provide a degree of protection against falling rain, sleet, and external ice formation. They shall meet rod entry, rain, external icing, and rust-resistance design tests. They are not intended to provide protection against conditions such as dust, internal condensation, or internal icing.

Type 4 Enclosures
Type 4 enclosures are intended for indoor or outdoor use primarily to provide a degree of protection against windblown dust and rain, splashing water and hose-directed water. They shall meet hose-down, dust, external icing, and rust-resistance design tests. They are not intended to provide protection against conditions such as internal condensation or internal icing.

Type 4X Enclosures
Type 4X enclosures are intended for indoor or outdoor use primarily to provide a degree of protection against corrosion, windblown dust and rain, splashing water, and hose-directed water. They shall meet the hose-down, dust, external icing, and corrosion-resistance design tests. They are not intended to provide protection against conditions such as internal condensation or internal icing.

(continued)

TABLE 11-1
NEMA Type Enclosures for Electrical Equipment (1000 Volts Maximum)
(continued)

Type 5 Enclosures

Type 5 enclosures are intended for indoor use primarily to provide a degree of protection against dust and falling dirt. They shall meet the dust and rust-resistance design tests. They are not intended to provide protection against conditions such as internal condensation.

Type 6 Enclosures

Type 6 enclosures are intended for indoor or outdoor use primarily to provide a degree of protection against the entry of water during occasional temporary submersion at a limited depth. They shall meet submersion, external icing, and rust-resistance design tests. They are not intended to provide protection against conditions such as internal condensation, internal icing, or corrosive environments.

Type 6P Enclosures

Type 6P enclosures are intended for indoor or outdoor use primarily to provide a degree of protection against the entry of water during prolonged submersion at a limited depth. They shall meet air pressure, external icing, and corrosion-resistance design tests. They are not intended to provide protection against conditions such as internal condensation or internal icing.

Type 11 Enclosures

Type 11 enclosures are intended for indoor use primarily to provide, by oil immersion, a degree of protection to enclosed equipment against the corrosive effects of liquids and gases. They shall meet drip and corrosion-resistance design tests. They are not intended to provide protection against conditions such as internal condensation or internal icing.

Type 12 Enclosures

Type 12 enclosures are intended for indoor use primarily to provide a degree of protection against dust, falling dirt, and dripping noncorrosive liquids. They shall meet drip, dust, and rust-resistance tests. They are not intended to provide protection against conditions such as internal condensation.

Type 12K Enclosures

Type 12K enclosure with knockouts are intended for indoor use primarily to provide a degree of protection against dust, falling dirt, and dripping noncorrosive liquids other than at knockouts. They shall meet drip, dust, and rust-resistance design tests. Knockouts are provided in the top and/ or bottom walls only. After installation, the knockout areas shall meet the environmental characteristics listed above. They are not intended to provide protection against conditions such as internal condensation.

(continued)

TABLE 11-1
NEMA Type Enclosures for Electrical Equipment (1000 Volts Maximum)
(continued)

Type 13 Enclosures
    Type 13 enclosures are intended for indoor use primarily to provide a degree of protection against dust, spraying of water, oil, and noncorrosive coolant. They shall meet oil exclusion and rust-resistance design tests. They are not intended to provide protection against conditions such as internal condensation.

Classified Location Enclosures

Type 7 Enclosures
    Type 7 enclosures are for indoor use in locations classified as Class I, Groups A, B, C, or D, as defined in the *National Electrical Code.*

Type 8 Enclosures
    Type 8 enclosures are for indoor or outdoor use in locations classified as Class II, Groups A, B, C, or D, as defined in the *National Electrical Code.*

Type 9 Enclosures
    Type 9 enclosures are intended for indoor use in locations classified as Class II, Groups E, F, or G, as defined in the *National Electrical Code.*

Type 10 Enclosures (MSHA)
    Type 10 enclosures shall be capable of meeting the requirements of the Mine Safety and Health Administration, 30 C.F.R., Part 18 (1978).

sition that allows for maximum air flow in the enclosure. Since the power supplies generate the most heat, they should not be mounted directly underneath another system component. Generally, the main power supply is mounted near the top of the enclosure, but some PC manufacturers use individual power supplies on each I/O rack and an internal power supply for the processor. In Figure 11-1 note the 4-inch horizontal gap between the I/O chassis to allow for adequate cooling air flow.

The temperature inside the control panel or enclosure must not exceed the maximum operating temperature listed in the manufacturer's installation manual (typically 120°F). If the temperature limit cannot be maintained by convection cooling, a fan or blower must be installed to help dissipate the heat. The fan or fans used will generally be equipped with filters to prevent dust, dirt and other airborne contaminants from entering the enclosure and affecting system components.

Maintenance Features   The system designer must include certain features in the design of the enclosure to reduce maintenance time and cost. One important consideration is the accessibility of equipment components and terminal connections. For example, the processor should be placed at working level for ease of operation or maintenance. If the processor and the system power supply are contained in a single unit, it should be placed near the top of the enclosure. However, if there is adequate space in the enclosure, it should be placed to improve operation or maintenance.

Another maintenance feature is that the control panel should have an ac power outlet strip so maintenance can plug in test equipment and a portable light if needed. If the panel is large, interior lighting should be installed to aid maintenance and operations personnel in troubleshooting. Both the ac power strip and the interior lighting should be on a separate ac circuit from the other system components.

In some applications, a gasketed Plexiglas™ window is used to allow for viewing of the processor status lights and/or I/O status indicators by operations and maintenance personnel. In most applications, the operator will check the status of process and I/O lights during each step in a process to make sure there are no abnormal operating conditions.

Panel Duct and Wiring Design   The wiring duct layout is determined by the types of signals used in the system design. It also depends on the placement of the I/O modules in the racks. The main consideration is to reduce the electrical noise caused by crosstalk between I/O signal lines.

All ac power wiring should be kept separated from low-level dc wires. If dc lines must cross ac signal or power lines, it should be done at right angles only. This routing practice minimizes the possibility of electrical interference.

Power Distribution Design   Most programmable controller manufacturers recommend that a power isolation transformer be installed between the ac power source and the PC equipment to provide for signal isolation from other equipment in the process area. A typical power distribution drawing is shown in Figure 11-2. In this application, the incoming ac lines (L1, L2, and L3) are 120 V ac, used for power devices, such as motors, heaters, or pump starters, in the system. There are three fuses on the incoming lines to protect against any overcurrent condition. In the drawing, the 120 V ac power lines L1 and L2 are connected to the primary of the step-down transformer. This transformer steps down or reduces the ac voltage to 120 volts and provides isolation for the programmable controller components from the outside electrical equipment.

The ac distribution circuit contains a "master control relay," labeled as CRM in the circuit diagram. This relay is used to stop the operation of the programmable controller or the controlled machine or process when any emergency stop (E-stop) push button is depressed by an operator. Any number of E-stop switches can be used in the ac distribution circuit to improve system safety.

It is also recommended that emergency stop circuits be designed into the system for every machine being directly controlled by the programmable controller. These circuits should be hard wired and totally independent of the programmable controller to provide for maximum safety in the control system.

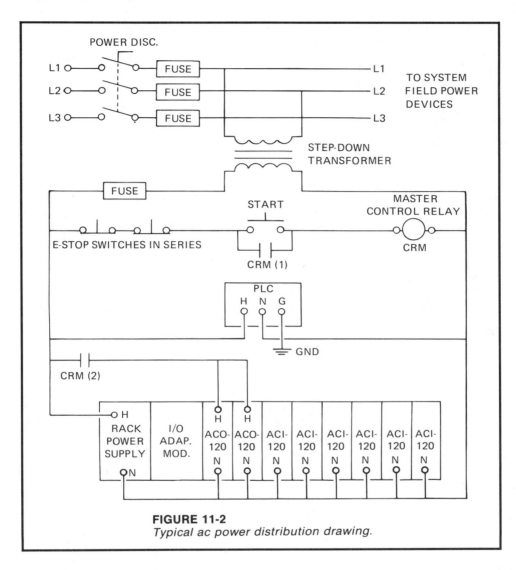

FIGURE 11-2
Typical ac power distribution drawing.

Programmable controllers are very reliable devices, but failure of the central processor can cause dangerous and erratic behavior of the control system. Therefore, the operator must be able to quickly and safely turn off process equipment and machinery by using E-stop switches that are placed in locations easily accessible to the operator.

Grounding Considerations   The reliability of any electrical control system is highly dependent on the proper design of system grounding. Correct electrical grounding design is also required to guarantee a safe electrical installation. When designing and installing any electrical equipment, the engineer should read and understand section 250 of the National Electrical Code (NEC). This section provides information on the wire color code, size, and type of conductor required as well as connection methods required for safe grounding of electrical equipment.

The code states that a ground must be permanent (i.e., no solder connections), continuous, and able to safely conduct the current in the system with minimal resistance. The ac hot wire must have black-colored insulation; the ac neutral wire insulation must be white; and the ground conductors must use green-colored insulation. It is common practice to use red insulation for positive signal wires and black-colored wires for negative dc signal lines, but it is not required by NEC.

The ground wire should be separated from the ac hot and ac neutral wires at the point of entry to the control panel. To minimize the ground wire length within the panel, the ground reference point should be located as close as possible to the point of entry of the panel supply line. All I/O racks, power supplies, processors, and other electrical devices in the system should be connected to a single ground bus in the control panel. Paint or other nonconductive materials should be scraped away from the area where an I/O rack makes contact with the panel. In addition to the ground connection made through

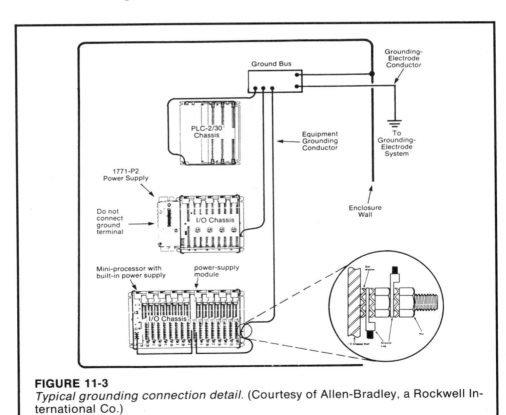

FIGURE 11-3
Typical grounding connection detail. (Courtesy of Allen-Bradley, a Rockwell International Co.)

the rack mounting bolts, a metal braid of the size recommended by the PC manufacturer should be used to connect each chassis and the panel at a single mounting bolt. Figure 11-3 shows a typical grounding connection detail.

I/O Module Installation and Wiring   An important part of the hardware design of a PC system involves the proper layout and wiring diagrams for the input/output module. The actual installation of the modules is a relatively simple procedure with almost all the programmable controllers on the market, since the installation crew simply plugs the modules into the I/O racks per the drawing provided by the system designer.

However, in some cases, intelligent modules (such as thermocouple, analog, communications, etc.) might require switch settings to properly configure the module. For example, a thermocouple (T/C) module might be designed to accept the various T/C types, such as J, K, T, S, etc., and the user sets toggle switches in the module to accept the T/C being used in the process. The design engineer must document the switch setting on the I/O drawings or in the installation instructions to guarantee that the modules are properly installed and configured. Figure 11-4 shows a typical T/C input module drawing with a switch setting table.

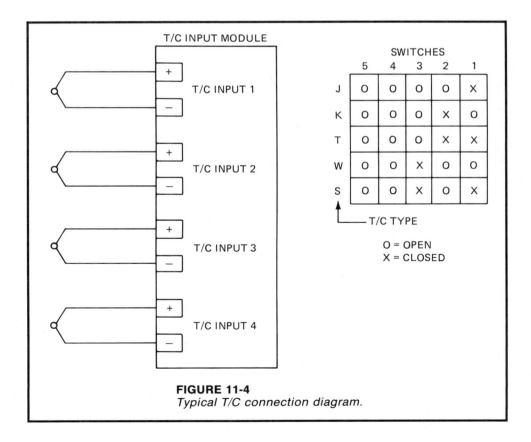

FIGURE 11-4
Typical T/C connection diagram.

I/O RACK NUMBER	SWITCH POSITION			
	4	3	2	1
01	CLOSED	CLOSED	CLOSED	OPEN
02	CLOSED	CLOSED	OPEN	CLOSED
03	CLOSED	CLOSED	OPEN	OPEN
04	CLOSED	OPEN	CLOSED	CLOSED
05	CLOSED	OPEN	CLOSED	OPEN
06	CLOSED	OPEN	OPEN	CLOSED
07	CLOSED	OPEN	OPEN	OPEN
08	OPEN	CLOSED	CLOSED	CLOSED

FIGURE 11-5
Typical rack number selection table.

Another important procedure required for installation of a programmable controller system is the selection of the I/O chassis number for each I/O rack. Many PC manufacturers use switch assemblies in the I/O rack to number the racks. A typical programmable controller I/O rack selection table is shown in Figure 11-5.

Control Panel Specification   In most cases the equipment enclosure for a programmable controller system is fabricated by an outside control panel vendor, so that an equipment specification will be required to cover all requirements of a system. A typical control panel specification is given in Table 11-2.

Equipment Layout Design   Proper control system equipment layout design can reduce installation costs and improve system reliability and maintainability.

In addition to the programmable controller components, the equipment layout must also take into account the other system components, such as field devices and instruments, power disconnect boxes, power transformer, and the location of process equipment and machines.

In general, placing the processor near the process equipment and using

Table 11-2
Typical Control Panel Specification

The control panel furnished under this specification shall be supplied complete with the instruments and equipment listed on the enclosed drawings, installed and electrically wired, and ready for wiring to field instruments and equipment.

The control panel shall conform to the following:
1. The control panel shall be fabricated with cold-rolled steel plate.
2. All miscellaneous items, such as wire raceways, terminal strips, electrical wire, etc., where possible, shall be made of fire-resistant materials.
3. The grounding bus bar and studs shall be pure copper metal.
4. A minimum of two 120-V ac, 60-Hz utility outlets shall be installed in panel.
5. An internal fluorescent light with a conveniently located ON/OFF switch shall be installed in the panel.
6. A circuit breaker panel with circuit breakers to accommodate all panel lighting, power supplies, instruments, I/O modules, processors, and any other loads listed on the system drawings shall be provided.
7. Each electric wire over 4 inches in length shall be identified at each end with a wire number per the electrical drawings for the system.
8. Terminal strips with screw-type connectors shall be used and no more than two wires shall be terminated on any single terminal.
9. Nameplates shall be made of laminated plastic with white letters engraved on a black background for both front and rear panel-mounted instruments and components, such as power supplies, transformers, I/O racks.
10. The ac wiring shall be separated from 4 to 20-mA dc current and digital signals by a minimum of 24 inches, and they must be wired to separate terminal strips.
11. All electrical wires and cables shall enter the control panel through the top. Sufficient space shall be provided to allow ac power cables entering the top of the panel to continue directly to the circuit breaker panel.
12. All wires and cables shall be routed and tied to provide clear access to all instruments and components for maintenance and removal of defective components.
13. All dc wires shall have a minimum insulation voltage rating of 600 volts ac, and all dc wires shall have a minimum insulation voltage rating of 300 volts dc.
14. All electrical conductors shall be copper of the correct wire size for the current carried with 98% conductivity, referenced to pure copper.
15. Wireways shall be attached securely to the control panel.
16. Metal surfaces with wires or cables passing through them shall be furnished with insulated polyethylene grommets to prevent damage to the conductors or cables.
17. All wire bundles or cables shall be clamped to the panel at all right angle turns.
18. All wires entering or leaving a wire bundle shall be tied to the bundle at the point of entering or leaving the main wire bundle.

remote I/O racks where possible will reduce wire and electrical conduit runs. It is possible that the cost of wiring and conduit installation will far exceed the programmable controller equipment costs if care is not taken in the equipment layout design.

The control panel should be placed in a position that allows the doors to be opened fully for easy access to wiring terminals and system components for maintenance and troubleshooting. The National Electric Code (NEC) requires that the panel doors open at least to 90°, and there must be a minimum of 36 inches of clearance from the rear of the panel to the nearest grounding surface or wall. Before starting the design of a control panel, the designers should read and understand section 110 of the National Electrical Code to avoid any code violations or safety problems.

An emergency disconnect switch should be mounted on or near the control panel in an easily accessible location.

If the location where the control panel will be placed contains equipment that generates excessive radio frequency interference (rfi) or electromagnetic interference (emi), the panel should be placed a reasonable distance from these sources. Examples of such sources include electric machinery, welding equipment, induction heating units, and electric motor starters.

System Start-Up and Testing   The first requirement for successful system start-up is to have a written system operational (SO) test procedure. This procedure should be written by the control system designer and carefully reviewed and approved by all parties involved in the project. A time-saving procedure is to functionally test the control system program using a process simulator before the system is installed on the process.

The SO test procedures will generally contain the following main sections: (1) visual inspection, (2) continuity test, (3) input signal testing, (4) testing of outputs, and (5) process operational testing.

A typical SO test procedure is given in Table 11-3.

Maintenance Practices   Programmable controller components are designed to be very reliable, but occasional repair is still required. System maintenance costs can be greatly reduced by using good design practices, complete documentation, and preventive maintenance programs.

Preventive Maintenance   A systematic preventive maintenance program will reduce the downtime of the control system. The preventive maintenance of programmable controller components is usually scheduled at the same time as the machine or process that is down for maintenance or repair. Normally, since programmable controller equipment is more reliable than some machinery or process equipment, it requires less frequent preventive maintenance operations.

It is very important in a preventive maintenance program to carefully check the various connectors in the system. It is estimated that 70% of the problems found in electronic or computer-based systems are caused by loose,

TABLE 11-3
Typical SO Test Procedure

I. Visual Inspection
  1. Verify that all system components are installed per the system drawing.
  2. Check I/O module location in equipment racks per I/O drawings.
  3. Inspect switch settings on all intelligent modules per drawings.
  4. Verify that all the system communication cables are correctly installed.
  5. Check that all input wires are correctly marked with wire numbers and terminated at correct points on input module.
  6. Check that all output wires are correctly marked with wire numbers and terminated at correct terminals on output module.
  7. Verify the power wiring is installed per the ac distribution drawing.
II. Continuity Check
  1. Use ohmmeter to verify that no ac wire is shorted to ground.
  2. Verify continuity of ac hot wiring.
  3. Check ac neutral wiring system.
  4. Check continuity of system grounds.
III. Input Wiring Check
  1. Place the programmable controller in test mode.
  2. Disable all output signals.
  3. Turn "on" ac system power and power to input modules.
  4. Verify the E-stop switch removes power from the system.
  5. Activate each input device, observe the corresponding address on the programming terminal, and verify that the indicator light on the input module is energized.
IV. Output Wiring Check
  1. Disconnect all output devices that might create a safety problem, such as motors, heaters, or control valves, etc.
  2. Place programmable controller in test mode.
  3. Apply power to the programmable controller and the output modules.
  4. Depress the E-stop push button and verify that all output signals are de-energized.
  5. Restart system and use the forcing function in the programming terminal to energize each output individually. Measure the signal at the output devices and verify that the output light on the module is energized for each output tested.
V. Operation Test
  1. Place processor in "Program" mode and turn on main power switch.
  2. Load the pre-tested control program into the programmable controller.
  3. Disable all outputs, select the run mode on the processor, and verify that the run light on the processor is activated.
  4. Check each rung of logic for proper operation by simulating the inputs and verifying on the programming terminal that the correct output is energized at the proper time or sequence in the program.
  5. Make any required changes to the control program.
  6. Enable output modules and place processor in "Run" mode.
  7. Test control system per process operating procedure.

dirty, or defective connectors. The connection to the I/O modules should be checked periodically to make sure no wiring has come loose. The seating of the I/O modules in the equipment rack should also be checked. If the system is located in process areas that have high vibration levels, the preventive maintenance check should be performed more often.

Excessive heat is another major cause of failure in programmable controller systems. Therefore, if a enclosure is cooled with fans, the filters used on the panel must be cleaned or replaced on a regular basis. It is important that dirt and dust is not allowed to build up on system components, because a dirt buildup on electronic components can reduce heat dissipation and cause an overheated condition in the system.

Electrical noise can cause erratic and dangerous operation of a PC system, so the maintenance department must check to make sure that equipment producing high levels of rfi or emi noise are not moved near the programmable controller equipment.

Maintenance personnel should also check to make sure that unnecessary items are not stored on or near the PC equipment. Leaving items such as tools, test equipment, drawings, or instruction manuals in the control panel can obstruct the air flow and cause heating problems.

An important maintenance practice is to stock spare parts for the system. This reduces downtime in the event of a component failure.

EXERCISES

11.1 Discuss the main reasons and advantages for placing programmable controller components in metal enclosures.

11.2 Discuss the need for designing control panels to meet the heat requirements of programmable controller system components.

11.3 List some important design practices used to reduce maintenance costs on programmable controller systems.

11.4 Design and draw a power distribution system for the programmable controller system shown below. Assume the system is mounted in a control panel and has a 6 ac outlet power strip, two cooling fans, and a single fluorescent light.

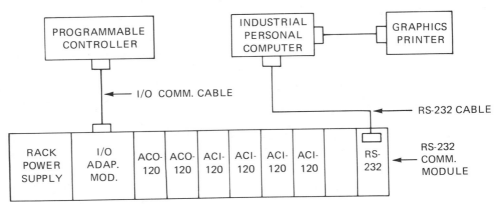

11.5 Discuss the importance of proper grounding techniques in programmable controller system design.

11.6 List the various phases of a typical system operational test procedure.

11.7 Discuss the important features of an effective preventive maintenance program for programmable controller systems.

BIBLIOGRAPHY

1. *Assembly and Installation Manual PLC-5 Family Programmable Controllers,* Allen-Bradley, a Rockwell International Company, Publication 1785-6.6.1, November, 1987.

2. Bryan, L. A., and Bryan E. A.; *Programmable Controllers, Theory and Implementation,* Industrial Text Co., 1988.

3. *Assembly and Installation Manual PLC-2/20, PLC-2/30 Programmable Controllers,* Allen-Bradley, a Rockwell International Company, Publication 1772-6.6.2, March, 1984.

4. *National Electrical Manufacturers Association (NEMA),* Enclosures for Electrical Equipment (1000 volts maximum), NEMA Standard Publication No. 250, 1979.

Index